T0305785

Uncertainty Quantification in Computational Science

Theory and Application in Fluids and Structural Mechanics

Uncertainty Quantification in Computational Science

Theory and Application in Fluids and Structural Mechanics

Editors

Sunetra Sarkar

Indian Institute of Technology Madras, India

Jeroen A S Witteveen

Center for Mathematics and Computer Science (CWI), The Netherlands

NEW JERSEY • LONDON • SINGAPORE • BEIJING • SHANGHAI • HONG KONG • TAIPEI • CHENNAI • TOKYO

Published by

World Scientific Publishing Co. Pte. Ltd.

5 Toh Tuck Link, Singapore 596224

USA office: 27 Warren Street, Suite 401-402, Hackensack, NJ 07601

UK office: 57 Shelton Street, Covent Garden, London WC2H 9HE

Library of Congress Cataloging-in-Publication Data
Names: Sarkar, Sunetra, author. | Witteveen, Jeroen A. S., author.
Title: Uncertainty quantification in computational science : theory and application in fluids and structural mechanics / Sunetra Sarkar (Indian Institute of Technology Madras, India), Jeroen A S Witteveen (Center for Mathematics and Computer Science (CWI), The Netherlands).
Description: [Hoboken] New Jersey : World Scientific, [2016]
Identifiers: LCCN 2016016299 | ISBN 9789814730570 (hc : alk. paper)
Subjects: LCSH: Computational fluid dynamics. | Aerodynamics--Data processing. | Structural analysis (Engineering)--Data processing. | Measurement uncertainty (Statistics)
Classification: LCC QC151.7 .S37 2016 | DDC 620.1/064--dc23
LC record available at https://lccn.loc.gov/2016016299

British Library Cataloguing-in-Publication Data
A catalogue record for this book is available from the British Library.

Desk Editor: V. Vishnu Mohan

Typeset by Stallion Press
Email: enquiries@stallionpress.com

Printed in Singapore

Contents

Preface vii

Acknowledgements ix

In Memoriam xi

Chapter 1: Uncertainty Quantification Applied to Aeroacoustics of Wall-Bounded Flows 1

Julien Christophe, Marlène Sanjose, Jeroen Witteveen and Stéphane Moreau

Chapter 2: Computational Simulations of Near-Ground Sound Propagation: Uncertainty Quantification and Sensitivity Analysis 33

Chris L. Pettit and D. Keith Wilson

Chapter 3: Efficient Uncertainty Analysis of Radiative Heating for Planetary Entry 63

Thomas K. West IV and Serhat Hosder

Chapter 4: Investigation About Uncertain Metastable Conditions in Cavitating Flows 91

Pietro Marco Congedo, Maria Giovanna Rodio, Gianluca Geraci, Gianluca Iaccarino and Remi Abgrall

Chapter 5: Uncertainty Quantification in Structural Engineering: Current Status and Computational Challenges 119

Sayan Gupta and Debraj Ghosh

Chapter 6: Uncertainty Quantification in Aeroelastic Problems 151

Sunetra Sarkar, Harshini Devathi and W. Dheelibun Remigius

Index 181

Preface

Over the last decade, research in the area of Uncertainty Quantification (UQ) has gathered a tremendous momentum, especially in the areas of fluids engineering and coupled structure-fluids systems. New algorithms and their more efficient adaptive variants have come up in rapid succession in recent years. The present volume gives an overview of the current state-of-the-art of UQ techniques with an emphasis on the family of spectral approaches like polynomial chaos expansion (PCE) — arguably the most popular technique in the community at present. I feel the application areas covered in this volume within the realm of fluid dynamics and structural interaction problems are quite diverse. Further, the chapters present a balanced mix of research and review articles. Research articles on the use of UQ in computationally challenging problems like aeroacoustics of rotating machines and planetary vehicles' reentry truly highlight the current level of computational aspirations in the community. On the other hand, the review article on the current status of UQ in near ground sound propagation problem introduces the reader to a very specialized and interesting area. The fourth and the last chapters on fluid applications talk about the propagation of uncertainty in a cavitation model, an important area in fluid dynamics with computationally involved simulation procedures. The last two chapters are reviews on the state of the art on UQ and sensitivity analysis as applicable to structural engineering and aeroelastic systems. The application of PCE covers a major part of both the chapters; its scope in estimating the parameters in an inverse problem is especially interesting. This volume tries to bring together some of the thriving research areas in uncertainty quantification in fluids, structures and coupled problems. It is hoped that readers from one community will be able to learn and borrow from the merits of the others.

Acknowledgements

Putting this volume together has been a mixed experience. The excitement of another opportunity to work together with my friend and colleague Jeroen turned to a shocking tragedy with his sudden and untimely passing. It was difficult to accept his passing away as Jeroen the researcher was so young and so full of promises. It was a huge blow for the project and momentarily it felt very difficult for me to go on. But as they say, time heals, and today we are at the completion stage of this book. It certainly gives some satisfaction but more importantly, I feel a sense of fulfilment at being able to keep my promise to Jeroen. And this would not have been possible without the strong support of all the contributors to this volume. In fact, it made all the difference. I would like to take this opportunity to thank all of them. Not only their help with the review process was crucial but also their kind words went a long way in getting back my energy and some of the lost enthusiasm, with which we started the book. Serhat, Stephane, Chris, Pietro, Debraj, a big thank you! You were great. I also would like to thank my students, Santhosh, Jithin and Thanusha for all their help, both technical and non-technical. Thanks are due to Steven and his team from World Scientific as well. They always have been very prompt with their suggestions and clarifications to every question I had. They were also wonderful with the cover design. Overall, this book gave me a marvellous opportunity to know and interact with colleagues whose work I value and learn from and I sincerely thank them for the experience. Finally, I convey my love and appreciation to Sayan and Tsampi, their trust in me is the reason I can go on.

Sunetra Sarkar
IIT Madras, India

In Memoriam

"In the summer of 2015, the international uncertainty quantification (UQ) and computational fluid dynamics (CFD) community lost a very brilliant, creative, and promising young scientist, Jeroen Witteveen. Jeroen was also a good friend of mine and it was a total shock to hear about his early and unexpected passing. Despite his young age, Jeroen made significant contributions to the UQ and CFD fields. Besides his intellectual and scientific excellence, I will always remember him as a very friendly, calm, and good gentleman. I will miss his friendship and our deep conversations on everything. I hope this book will help to keep Jeroen's memory alive."

— Serhat Hosder
Missouri University of Science and Technology, USA

"Jeroen's passing was a terrible shock. I first met him at an AIAA conference when he was presenting his doctoral research. His work overlapped substantially with some research I had done. After finishing his doctorate, Jeroen warmly provided me with a hardcopy of his dissertation. In the subsequent years I was very happy to see him successfully climbing the ladder as a productive scholar and a strong proponent of our uncertainty quantification research. He was a genuinely good person who deserved only good things from life. He did great research, but I will remember him also as a kind-hearted person."

— Chris Pettit
United States Naval Academy, Annapolis, USA

"I have known Jeroen six years ago during the CTR Summer Program in 2010. Our collaboration was extremely fruitful and pleasant. Nevertheless, it was rare to exchange with him after work. Afterwards, we have started a work collaboration all over these years. Last time that I have seen him was in Paris at INRIA, spending an entire afternoon to think about future common actions concerning robust design. This brainstorming has permitted to get a funding for a joint PhD, that unfortunately has never started.

I think that the first time we have really started to interact as friends and not only as colleagues was during a common stay in Brussels two years ago. At that time, I have discovered a very funny person with a great sense of humour. It was a dinner that I will remember forever when thinking about him. I was lucky to know him, and will remember him forever as a great researcher, and a clean and pleasant person."

— Pietro Marco Congedo
INRIA, France

"Our contributed chapter in this volume is dedicated to Jeroen as he was one of the main instigator of the present UQ research and the chapter was the result of two fruitful and enjoyable collaborations during two consecutive CTR Summer Programs at Stanford. His sudden death came as a shock and Jeroen should be remembered as a deep and thoughtful researcher with whom it was marvellous working with."

— Stephane Moreau
Université de Sherbrook, Canada

"Jeroen's passing will be felt deeply by many; with his departure, the uncertainty quantification and fluid dynamics community undoubtedly lost one of their rising stars. He was well respected in the community for his commendable accomplishments within the span of a rather short-lived research career. I have known him since my postdoctoral days and had many wonderful opportunities to work with him. He had the great gift of superior imagination to turn every problem to a 'meaty' one and I have learnt a lot from him. He also had an open mind and a friendly personality; a quality which won him many friends all over. Colleagues will always remember him as an exceptionally talented and motivated researcher whose very promising career ended prematurely in its prime. He was also a good person at heart, and always ready to discuss any interesting topic under the sun with the ability to turn it more interesting in the process. This is the

Jeroen I am going to miss most and of course all the future opportunities of meeting him in a conference or meeting at some unfamiliar coordinates of this globe. Stay well Jeroen, till we plan again."

— Sunetra Sarkar
Indian Institute of Technology Madras, Chennai, India

Chapter 1

Uncertainty Quantification Applied to Aeroacoustics of Wall-Bounded Flows

Julien Christophe*, Marlène Sanjose†, Jeroen Witteveen‡ and Stéphane Moreau†

**Institut von Karman, 72 Chaussée de Waterloo B-1640 Rhode-Saint-Genese, Belgium, julien.jchristophe@vki.ac.be*

†*Université de Sherbrooke, 2500 boulevard de l'université Sherbrooke, QC, Canada, J1K2R1*

‡*Center for Mathematics and Computer Science (CWI) Amsterdam, The Netherlands*

The uncertainty quantification (UQ) related to the self-noise prediction based on a Reynolds-Averaged Navier-Stokes (RANS) flow computation of a low-subsonic axial fan has been achieved. As the methodology used for fan noise prediction is based on airfoil theories, the uncertainty quantification of a low-speed Controlled-Diffusion (CD) airfoil has been first considered. For both applications, deterministic incompressible flow solvers are coupled with a non-intrusive stochastic collocation method, found to be two orders-of-magnitude more efficient than a classical Monte Carlo simulation for the same accuracy. In the case of airfoil UQ, the effective flow angle is used as a random variable. Two wall-pressure reconstruction models are used to obtain necessary inputs of Amiet's trailing-edge noise model: Rozenberg's model has larger uncertainties at high frequencies because of the uncertainty on the wall-shear stress parameter required in the method, and Panton & Linebarger's model is less accurate at low frequencies because of the slow statistical convergence of the integration involved in the model. Similar behaviours are observed in the fan UQ involving the volume flow-rate and the rotational speed as random variables. The stochastic mean sound spectra are found to be dominated by the tip strip and compare well with experimental data. Larger uncertainties are seen in the hub and tip regions, where large flow detachment and recirculation appear. The known uncertainties on flow rate yield larger uncertainties on sound than those on rotational speed.

1. Introduction

In modern rotating machines, significant effort has been put to reduce annoying tonal noise, either by passive devices or by active noise control. The next challenge is then to reduce the broadband contribution to decrease the overall noise level and meet increasingly stringent environmental noise regulations. A key source of broadband noise is the trailing-edge noise or self-noise, caused by the scattering of boundary-layer pressure fluctuations into acoustic waves at the trailing edge of any lifting surface. In the absence of any interaction noise source, it represents the dominant source of noise generated by rotating machines such as low-speed fans, high-speed turboengines,[1] wind turbines[2] and other high-lift devices.[3] However, an accurate prediction of the sound by a rotating system still remains a daunting task by a direct computation (a compressible Large Eddy Simulation (LES) for instance). A hybrid approach combining a near-field turbulent flow simulation and an acoustic analogy for the sound propagation in the far-field is therefore preferred. Such a method has been thoroughly validated for broadband noise prediction on multiple airfoils in various flow conditions including blowing.[4–13] The computational cost associated with unsteady turbulent flow simulations still limits most numerical studies to simpler geometries such as airfoils,[8] even with sophisticated non-boundary-conforming methods such as the Lattice Boltzmann method and Immersed Boundary method,[4,14] or the use of unstructured grid topologies.[5] To meet industrial design constraints of rotating machines, approaches that model the pressure and velocity fluctuations needed for an acoustic analogy from steady RANS are often used.[15–18] These methods add further levels of modeling and the associated uncertainties grow, which may make the final acoustic prediction of fan broadband self-noise inaccurate and unreliable.

To illustrate this point, some of the aleatory uncertainties associated with the prediction of trailing-edge noise are considered, first for the wall-bounded canonical case of airfoil noise as measured in an open-jet wind tunnel,[6,19] and then for the actual complex case of a low-speed automotive engine cooling fan as tested in a reverberant wind tunnel.[20] For the former the uncertainty propagation from uncertain inlet velocity profiles caused by inaccurate knowledge of the jet deflection induced by the airfoil is studied. For the latter the uncertainty propagation from uncertain operating conditions (both volume flow-rate through the fan and rotational speed) mainly caused by industrial process issues is considered. In doing so, two representative wall-pressure models derived from steady RANS are dealt with, and compared with direct unsteady LES predictions of the trailing-edge noise for the airfoil case. The sensitivity of the RANS and LES solutions to inlet conditions and the uncertainty introduced by wall-pressure models coupled with RANS on the prediction of the noise sources and the far-field pressure are then assessed. The difference is made here between uncertainties in the physical inputs to the problem (aleatoric uncertainties) and the constants or variables of the model used to solve for the flow, for instance all constants used in the turbulence modeling, or for the far-field noise (epistemic uncertainties).[21]

The methodology for uncertainty quantification based on simulations of either a standard experimental setup for trailing-edge airfoil noise or a wall-mounted fan in a standard interface is presented in Sec. 2. The present stochastic approach is outlined in Sec. 3. The sound prediction methods are then presented in Sec. 4. The next two sections, Secs. 5 and 6, show the uncertainty quantification for the airfoil and the fan cases respectively. For both examples the deterministic flow simulations are briefly outlined, the random variables are described, and the stochastic aerodynamic and acoustic results are compared with the available experimental data. Conclusions are finally drawn in Sec. 7.

2. Uncertainty Quantification Methodology

The methods involved in the fan noise prediction rely on airfoil aeroacoustic models,[22] and the uncertainty related to such models is first assessed. The approaches to compute airfoil or fan trailing-edge noise are illustrated in Fig. 1.1. The directions of the arrows outline the logical sequence of the method. Starting from the fan blade geometry (grey square box), two different methodologies based on similar computational methods are used to study trailing-edge noise in the case of airfoil or fan applications.

On the one hand, a mid-span cut of an automotive cooling fan is performed to obtain a two-dimensional profile, and used to study uncertainty for airfoil trailing-edge noise. As in previous studies[4–13] the same validated numerical method for predicting trailing-edge noise is used. A computation of the complete experimental setup of the large anechoic wind tunnel

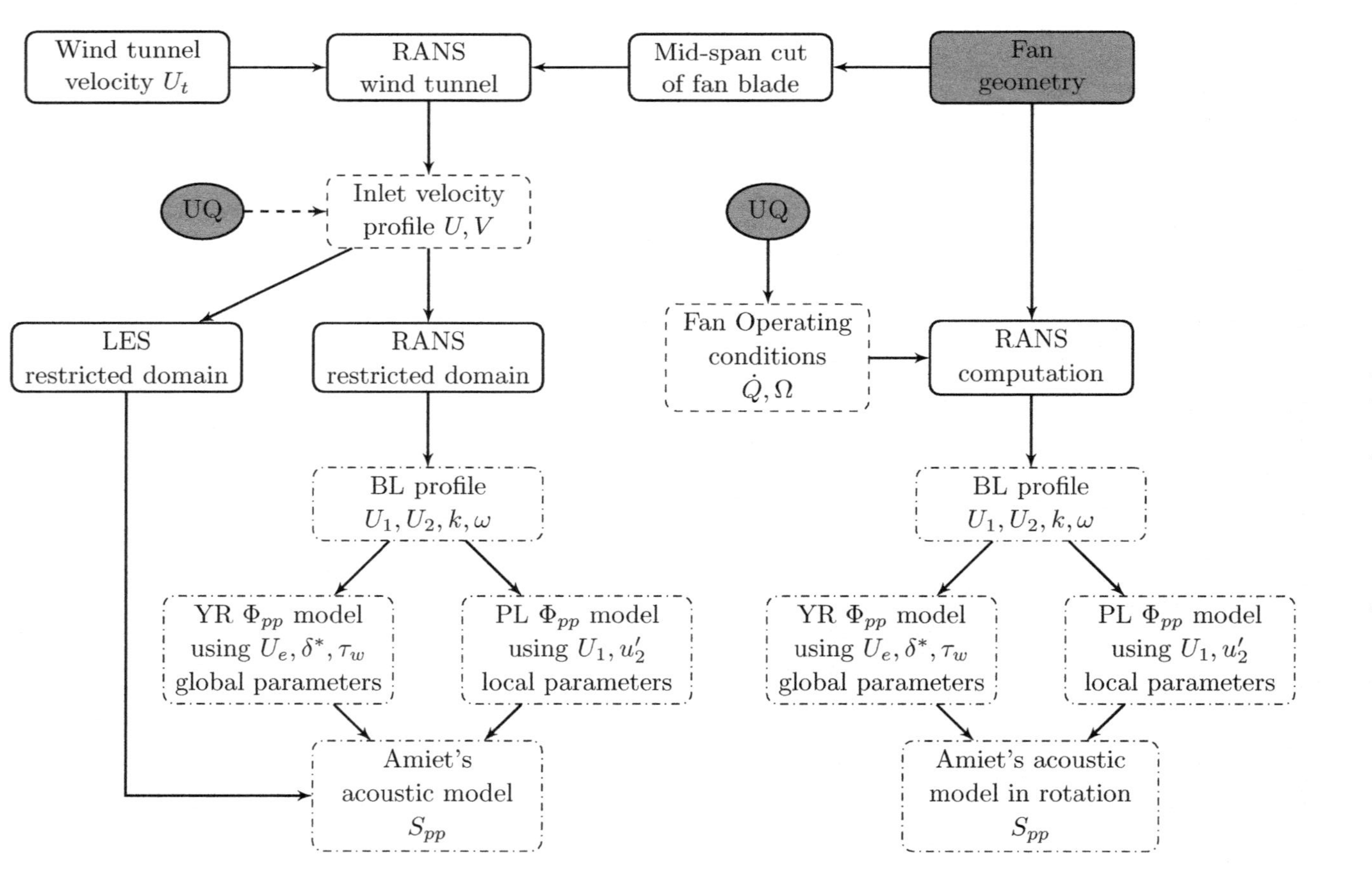

Fig. 1.1. Uncertainty quantification methodology: the solid-line rectangular boxes refer to the different computational steps in the hybrid methodology, the dashed-line rectangular boxes denote the main inputs and the dash-dotted-line boxes denote the outputs at each step and the arrows indicate the various possible workflows to yield the final acoustic pressure. The round box refers to the point where uncertainty quantification is introduced.

in Ecole Centrale de Lyon (LWT), including the nozzle and part of the anechoic chamber is first done in order to capture the strong jet-airfoil interaction and its impact on airfoil loading.[23] The input parameters for this computation, termed "wind tunnel" in Fig. 1.1, are the wind tunnel velocity U_t, air density ρ, and kinematic viscosity ν. The computational setup, defined by the nozzle and the airfoil geometries as well as the position of the airfoil in the wind tunnel and its geometrical angle of attack α_w with respect to the nozzle axis,[23] is directly taken from measurements on the experimental setup, and is therefore considered as a minor source of possible error on the final sound prediction results. Similarly, the uncertainty on the air density ρ and kinematic viscosity ν, obtained from the experiments ambient conditions, are also neglected. Once a value for the wind tunnel velocity U_t is selected, a RANS computation is run on the complete setup and boundary conditions are extracted (U and V profiles) for a smaller domain, termed "restricted domain" in Fig. 1.1, embedded in the jet potential core as shown in Fig. 1.2(a). The final sound prediction is obtained by two different procedures, both producing a wall-pressure frequency spectrum Φ_{pp} used in Amiet's theory to predict the far-field sound spectrum S_{pp}. In the first approach, which is simpler and more intuitive but also more computationally expensive, an unsteady LES on the restricted domain with the steady velocity profiles extracted from RANS is used to obtain a direct prediction of the trailing-edge wall-pressure spectrum. In the clean inflow of the jet potential core, no perturbations are introduced at the inlet of the computational domain. The second approach, which is less expensive but requires more modelling, uses steady RANS computations on a two-dimensional slice of the restricted domain, with the same boundary condition profiles as for the LES. From this RANS computation, the primitive variables (U_1 and U_2, velocity components parallel and perpendicular to the wall surface respectively, and k and ω or ϵ depending on the RANS turbulence model) are extracted through a boundary-layer profile at the trailing edge of the airfoil. Those variables are then used in the two wall-pressure models investigated in the present study. The first model is that of Rozenberg *et al.*[17] (termed YR), who proposed a model only based on global boundary-layer parameters from the boundary-layer profile, namely the external velocity U_e, a boundary-layer thickness δ or instead the displacement thickness δ^*, and the wall shear stress τ_w. The second model by Panton & Linebarger[24] (termed PL) uses local parameters i.e. the wall parallel velocity profile $U_1(y)$ and the wall perpendicular velocity fluctuation profile $u_2'(y)$. For validation purposes, LES output quantities would be also used in the wall-pressure models, as illustrated in a previous study.[18] The uncertainty is introduced in both velocity components (U and V), as shown in Fig. 1.1 by the UQ box, correlated via the non-uniform inlet flow angle at the inlet boundary condition on the restricted computational domain as described in Sec. 5. It not only represents the actual experimental uncertainty but also the uncertainty in the prediction of the flow deflection induced by the airfoil-jet interaction by various turbulence models in the RANS simulation of the whole wind tunnel.[23] The introduction of uncertainty is investigated through approaches, using LES to directly obtain the trailing-edge wall-pressure spectrum and using RANS and wall-pressure spectrum models. A direct comparison of all the methods of prediction of wall-pressure spectrum can then be achieved in terms of mean values and uncertainty bars. These uncertainty bars refer to the numerical results (either RANS or LES) only, and they come from the propagation of all the considered uncertainty introduced on the inlet velocity components to the final aerodynamic and aeroacoustic numerical predictions. When possible, the different steps of the methodology are compared with available experimental measurements and the RANS intermediate results are compared with data obtained directly from the LES. Other possibilities to introduce uncertainty, not shown in Fig. 1.1 and not applied in the present work, would be to perturb the wind tunnel velocity U_t used in the computation of the complete wind tunnel setup, leading to negligible effects on the far-field sound spectrum, or on the boundary-layer parameters used in the wall-pressure reconstruction models, which does not allow a direct comparison between LES and RANS prediction methods.

On the other hand, the approach to UQ to compute fan trailing-edge noise is also illustrated in Fig. 1.1. As in Christophe *et al.*,[25] a RANS computation of flow in a blade passage of the fan is first performed. The input parameters for this computation are the volume flow rate $\dot{Q}$ and the rotational speed Ω. Both parameters are considered as main parameters for the uncertainty quantification on fan operating conditions (solid circle in Fig. 1.1). These uncertainties actually correspond to the dispersion found in engine cooling modules induced by the process scattering of electrical motors (scatter in rotational speed) and heat exchangers (scatter in flow rate). In a blade-to-blade plane at a given radius, the speed triangles yield similar velocity components as in the airfoil case, which relate both uncertainty quantification studies. The computational setup, defined by the fan and tip gap are directly taken from the design of the experimental setup, and is therefore considered as a minor source of possible error on the final sound prediction results. From this RANS computation, the primitive variables are extracted through a boundary-layer profile at the trailing-edge of the fan blade, at five radial positions along the blade span. Those variables are then used to reconstruct wall-pressure fluctuation spectra Φ_{pp}, using similar reconstructions methods as for the airfoil application. Finally, the latter power spectral densities provide the far-field sound spectrum S_{pp} in the extended Amiet's theory,[26] applied in rotation as explained below in Sec. 4.

3. Stochastic Method

Classical methods for stochastic differential equations rely on Monte Carlo (MC) simulations that prescribe ensemble random inputs to these equations and then collect their ensemble solution realizations. They only require running a deterministic solver repetitively and do not depend strongly on the stochastic dimensionality of the problem; however, they suffer from a slow convergence rate and require a large number of samples, which is prohibitive for turbulent flow realizations. An example of this method is presented below in the airfoil case. Alternative methods are sensitivity methods based on the moments, perturbation methods where all stochastic variables are expanded in Taylor expansions around their mean,[27,28] stochastic collocation methods, and spectral or Galerkin projection methods.[29–31] The first two methods either strongly depend on the modeling assumptions or are limited to small variations. As existing flow solvers are used, non-intrusive stochastic methods are preferred, which means that projection is not applied to the Navier-Stokes equations directly, but rather to its inputs and outputs. Among the various projection methods, the Stochastic Collocation expansion (SC) is selected as it has proved its efficiency for flow simulations in various regimes.[32–34]

In this framework[32–34] the vector of random input parameters is given by $\vec{\xi} \in \Xi$, with Ξ the underlying parameter space and probability density $f_\xi(\vec{\xi})$. The objective of UQ is to compute the probability distribution and the moments μ_{X_i} of output of interest $X(\vec{\xi})$ defined as

$$\mu_{X_i} = \int_\Xi X(\vec{\xi})^i f_\xi(\vec{\xi}) \mathrm{d}\vec{\xi}. \tag{1.1}$$

The statistical moments such as the mean and standard deviation can then be calculated. The SC method solves this problem by approximating the functional relationship between the outputs of interest $X(\vec{\xi})$ and the random inputs $\vec{\xi}$ using an expansion in Lagrange polynomials. This results in the one-dimensional case with one uncertain input parameter ξ,

$$X(\xi)^i \approx \sum_{j=1}^{N} X(\xi_j)^i L_j(\xi), \tag{1.2}$$

with $L_j(\xi)$ the Lagrange polynomials defined as

$$L_j(\xi) = \prod_{\substack{k=1 \\ k \neq j}}^{N} \frac{\xi - \xi_k}{\xi_j - \xi_k}, \tag{1.3}$$

such that $L_j(\xi_k) = \delta_{jk}$ with δ_{jk} the Kronecker delta. The coefficients $X(\xi_j)$ are solutions of the aeroacoustic problem for the realization ξ_j for the random input parameter ξ.

The deterministic sampling points ξ_j in the SC method are usually chosen to be numerical quadrature points, because they are accurate for computing the integrals (Eq. (1.1)). Standard Gauss-quadrature requires recomputing each modal coefficient in Eq. (1.2) each time N is increased, which is prohibitive for LES. In the present study, we use the Clenshaw–Curtis[35] (CC) quadrature points for the sampling points ξ_j given by $\xi \in [-1, 1]$

$$\xi_j = -\cos\left(\frac{\pi(j-1)}{N-1}\right). \tag{1.4}$$

The points are linearly scaled for input parameters on other ranges. These points are nested in the sense that all points ξ_j are re-used when increasing the quadrature level l, allowing to keep the abscissae when N is increased, with $N = 2^l + 1$ for $l > 0$ and $N(0) = 1$. Substituting Eq. (1.2) into Eq. (1.1) leads to the approximation

$$\mu_{X_i} \approx \sum_{j=1}^{N} X(\xi_j)^i w_j, \tag{1.5}$$

with integration weights w_j

$$w_j = \int_{\Xi} L_j(\xi) f_\xi(\xi) \mathrm{d}\xi. \tag{1.6}$$

The SC method can be extended to multiple independent input uncertainties using the following isotropic full tensor product extension into n_ξ dimensions $\vec{\xi} = \{\xi_1, \ldots, \xi_{n_\xi}\}$

$$\mu_{X_i} = \sum_{j_1=1}^{N} \cdots \sum_{j_{n_\xi}=1}^{N} X(\xi_{j_1}, \ldots, \xi_{j_{n_\xi}})^i w_{j_1} \cdots w_{j_{n_\xi}}. \tag{1.7}$$

The stochastic input $\vec{\xi}$ or output $X(\vec{\xi})$ can also depend on other parameters such as spatial coordinates (velocity profiles and maps, pressure coefficient) or frequency (wall-pressure and far-field acoustic pressure power spectral density (PSD)). The final value of the stochastic expansion of the analyzed variable (maximum value of N) is termed P.

4. Sound Prediction Methods

As shown by Roger & Moreau,[26] the trailing-edge noise can be obtained by iteratively solving scattering problems at the airfoil edges. The main trailing-edge scattering obtained by Amiet[36] has been corrected by a leading-edge back-scattering contribution that fully accounts for the finite chord length. The random predicted sound field at a given observer location $\mathbf{x} = (x_1, x_2, x_3)$ and for a given radian frequency ω_f (or wavenumber k_f) then reads

$$S_{pp}(\mathbf{x}, \omega_f) = \left(\frac{\omega_f C x_3}{2\pi c_0 S_0^2}\right)^2 \frac{L}{2} |\mathcal{L}|^2 \Phi_{pp}(\omega_f) l_y(\omega_f), \tag{1.8}$$

where Φ_{pp} is the wall-pressure power spectral density and l_y the spanwise correlation length near the trailing edge assumed to be deterministic. The radiation integral $\mathcal{L}$ of which parameters are both the free stream velocity U_∞ and the convection speed can be found in Roger & Moreau.[26]

In case of rotation, the far-field noise PSD of a low solidity fan with B independent blades is given by an integration over all possible azimuthal positions[22] of the single airfoil formulation (1.8)

$$S_{pp}(\mathbf{X}, \omega_f) = \frac{B}{2\pi} \int_0^{2\pi} \left(\frac{\omega_e(\Psi)}{\omega_f}\right)^2 S_{pp}^{\Psi}(\mathbf{x}, \omega_e) d\Psi. \tag{1.9}$$

The factor $\omega_e(\Psi)/\omega_f$ accounts for Doppler effects caused by the rotation (ω_e emission angular frequency). The fan noise predictions therefore rely on a strip theory combined with an acoustic analogy,[37] originally developed by Schlinker & Amiet[22] for helicopter rotor's trailing-edge noise, and extended to finite chord lengths and general three-dimensional gusts by Roger & Moreau,[26] and applied to low speed fans by Moreau & Roger.[38] Recently Sinayoko *et al.* has shown that this approximation of locally translating airfoils is accurate except at transonic speeds.[39] In order to take into account the variation of the flow along the blade span, the latter is split into 5 equal segments from hub to tip as in Christophe *et al.*,[25] and the total radiated sound is then the summation of the sound emitted by each airfoil strip.

When an unsteady LES computation is used, the wall-pressure spectra Φ_{pp} at the trailing edge is directly extracted from the simulations. In the RANS simulations, all variables are time-averaged and the wall-pressure fluctuations are reconstructed from the mean flow. Two such models are considered.

The YR model only uses integral boundary-layer parameters and reads

$$\frac{\Phi_{pp}(\omega_f)\,U_e}{\tau_w^2\,\delta} = \frac{0.78\,(1.8\,\Pi\,\beta_C + 6)\,\left(\frac{\omega_f\delta}{U_e}\right)^2}{\left[\left(\frac{\omega_f\delta}{U_e}\right)^{0.75} + 0.5\right]^{3.7} + \left[1.1\left(\frac{\omega_f\delta}{U_e}\right)\right]^7}, \tag{1.10}$$

where Clauser's parameter is $\beta_C = (\Theta/\tau_w)(dp/dx)$ and Coles' parameter Π is given by the implicit law of the wake:[40] $2\Pi - \ln(1+\Pi) = \frac{\kappa U_e}{u_\tau} - \ln\left(\frac{\delta^* U_e}{\nu}\right) - \kappa C - \ln\kappa$ (Θ the momentum thickness, κ the von Kármán constant and C a constant in the log law). For Eq. (1.10), the original expression[41] based on δ has been preferred to the formulation recently proposed by Rozenberg *et al.*[17] based on δ^*, as it provides a better agreement with the experimental data for the CD airfoil.[8,23,42] As explained by Rozenberg *et al.*,[17] the friction velocity u_τ is obtained from extended Clauser plots rather than direct upwind finite-difference estimates. Similarly the wake law is also verified graphically by comparing the model with the measured dimensionless velocity log plots (u^+, y^+).

Using the PL model,[24] Remmler *et al.*[18] derived an expression for the wall-pressure spectrum

$$\Phi_{pp}(\omega_f) = 8\rho^2 \iiint_0^\infty \frac{k_f^1(\omega_f)^2}{k_f(\omega_f)^2} \exp^{-k_f(\omega_f)(y+\hat{y})} \ldots$$
$$S_{22}(y,\hat{y},\omega_f)\frac{\partial U_1}{\partial y}\frac{\partial U_1}{\partial \hat{y}} dy d\hat{y} dk_f^3 \tag{1.11}$$

with the energy spectrum of the vertical velocity fluctuations:

$$S_{22}(y,\hat{y},\omega_f) = \frac{\bar{u_2'}(y)\bar{u_2'}(\hat{y})}{\pi^2}\Lambda^2 \iint_0^\infty R_{22}\ldots$$
$$\cos(\alpha\, k_f^1(\omega_f) r_1)\cos(\alpha k_f^3 r_3) dr_1 dr_3. \tag{1.12}$$

The model therefore uses the wall parallel velocity profile U_1 and the wall perpendicular velocity fluctuation profile u_2'. Following the procedure outlined by Remmler *et al.*,[18] both velocities and the velocity correlation length scale Λ are calculated from the RANS outputs and the velocity correlation

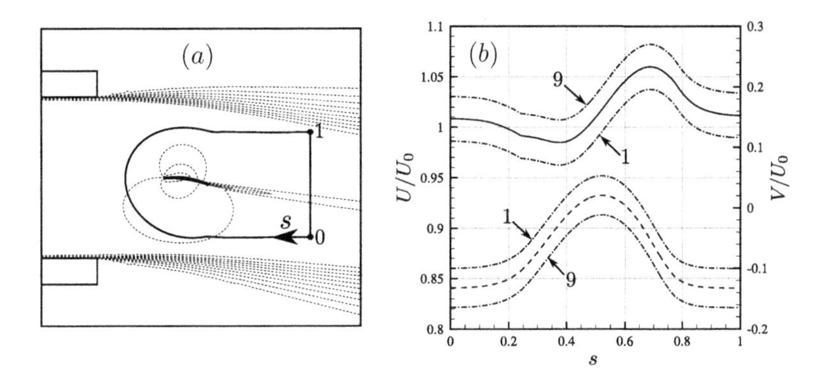

Fig. 1.2. (a) Parameterization of inlet boundary condition (solid line) and velocity magnitude contours (dash line), levels 0–1.7 with increment 0.1. (b) Parametric inlet velocity profiles: (plain) streamwise velocity U, (dash) crosswise velocity V and (dash-dot) uncertainty bounds around inlet profiles.

function R_{22} and the scale anisotropy factor α are modelled. No quadratures are used to calculate the quintuple integral in Eq. (1.11) as they would require prohibitive memory. The integration is performed with a Monte Carlo method using importance sampling for enhancing convergence. More details on this model can be found in Remmler *et al.*[18]

5. Controlled-Diffusion Airfoil

5.1. *Deterministic flow simulations*

The present computations consider the Controlled-Diffusion (CD) airfoil studied previously.[4–9] This profile corresponds to the mid-span section of a typical automotive cooling fan with 9 symmetric blades, the H380EC1 fan blade considered in Sec 6.[43] It has a 4% relative thickness and a leading-edge camber angle of 12°. The airfoil chord length is $C = 0.1356$ m. It is set at a geometrical angle of attack of $\alpha_w = 8°$. The reference velocity is $U_0 = 16$ m/s, yielding a Reynolds number based on the airfoil chord length $\mathrm{Re}_C=1.6\times10^5$. Further details on the experimental setup and flow conditions can be found in Moreau *et al.*[23]

The RANS computations are obtained with the Ansys Fluent 12 solver using the Shear-Stress-Transport (SST) $k-\omega$ turbulence model.[44] Unlike several $k-\epsilon$ models, the SST model was shown to properly capture the laminar recirculation bubble on this airfoil.[23] Second-order schemes are used for spatial discretization of all variables. The RANS computations use a no-slip boundary condition on the airfoil surface, a convective outflow boundary condition at the exit plane, and velocities (U and V) from the wind tunnel computation at the inlet. RANS simulations were run until a convergence to machine accuracy was reached.

The LES are based on the spatially filtered, incompressible Navier-Stokes equations with the dynamic subgrid-scale model.[45,46] These equations are solved using energy-conserving non-dissipative central difference schemes for spatial discretization and the fractional-step method for time advancement,[47] leading to the control-volume solver CDP originally developed by Mahesh *et al.*,[48] for hybrid unstructured grids. Details on the numerical schemes for the current unstructured solver can be found in Ham & Iaccarino.[49] The mesh taken from Moreau *et al.*[5] and labelled *CDP-B* is similar to the structured mesh of Wang *et al.*[8] but a spanwise grid coarsening proportional to the distance from the airfoil is performed to reduce the overall number of nodes to 1.5 million. The spanwise extent of the computational domain is again taken to be 10% of the chord length, which is enough to capture the experimental correlation length for this flow condition.[8,42] *CDP-B* LES results were shown to compare very well with both experimental wall-pressure and wake-velocity data.[5] Except in the vicinity of the leading edge, the near-wall grid resolution on the suction side is $\Delta x^+ \leq 34$, $\Delta y^+ \leq 1$, and $\Delta z^+ \leq 20$ in wall units, which is adequate for LES.[50] On the pressure side, the resolution is coarser because the boundary layer is laminar. The same inflow/outflow conditions as for the RANS are used. Periodic boundary conditions are applied in the spanwise direction. For the LES simulations that were run, a dimensionless time step of $2.5 \cdot 10^{-4}$, normalized by the chord C and the reference velocity U_0, is used leading to a CFL smaller than one in the whole domain. After a steady state was reached, airfoil trailing-edge surface pressure and flow statistics were collected for 50 flow-through times to simultaneously yield a sufficiently high spectral resolution (10 kHz) and a good statistical convergence of the low frequency spectral components (down to 100 Hz).

5.2. *Characterization of the random variables* $\vec{\xi}$

Figure 1.2(a) shows iso-contours of velocity magnitude from a RANS computation on the full wind tunnel configuration (LWT). The wind tunnel velocity was set to the nominal value $U_t/U_0 = 1$. On the same figure, the boundary of the restricted computational domain is represented, showing that the restricted domain is fully embedded in the inviscid jet potential core between the nozzle shear layers. The velocity components (U and V) from this simulation are interpolated on the boundaries of the restricted domain and are used as inlet conditions of the following steps for the methodology described in Sec. 2. The averaging along the curvilinear abscissa s of the inlet velocity profiles shown in Fig. 1.2 (b) provides an inlet velocity vector as shown in the sketch in Fig. 1.3. A corresponding inlet flow angle relative to the airfoil chord α_e (see definition in Fig. 1.3) can be computed and is found as $4°$ for the reference numerical deterministic case having a geometrical angle of attack of $8°$. The physical variations observed in the experimental flow measurements are taken into account by selecting a 2.5% uncertainty bound on the streamwise velocity U and a 10% uncertainty bound on the crosswise velocity V that include the short-scale fluctuations seen in the experimental profiles. Further details on the selection of the uncertainty bounds can be found in Christophe *et al.*[6] These conservative uncertainty ranges for both velocity components are also consistent with the observed variations found in the RANS simulations with different turbulence models.[23] The corresponding bounds of inlet flow angle α_e are $[6°, 2°]$, and the relative angular variation $\Delta\alpha_e$ with respect to the above reference case ($\alpha_e = 4°$) or simulation #5 is then $[2°, -2°]$ (see Table 1.1). The same computational mesh is used for all computations as only the inlet flow angle α_e is varied through the boundary conditions while the geometrical angle of attack α_w is kept constant. Furthermore, due to the small variations of the inlet flow angle, the airfoil wake angle remains close to the reference case for all computations and therefore the same grid extent can be used for all computations without having any interaction with the domain boundary conditions. The upper and lower bounds of the velocity profiles are shown in Fig. 1.2, together with the deterministic profiles. Both components U and V, and consequently the inlet flow angle α_e, are assumed to be random variables ($\vec{\xi}(\gamma) \equiv (U(\gamma), V(\gamma))$ or $\xi(\gamma) \equiv \alpha_e(\gamma)$) with uniform distribution ζ of the random perturbation parameter γ within their interval of variation. From those bounds, a set of 17 velocity inlet profiles (P=17 in Eq. (1.2)) are determined using a Clenshaw-Curtis quadrature.[35] A total of 17 corresponding RANS computations are run showing a sufficient convergence of the stochastic collocation method for 9 samples, as explained in Sec. 5.3.2, and only 9 LES computations are then run (P=9 in Eq. (1.2)). Table 1.1 summarizes the inlet flow angle variation for the 9 samples with γ. MC simulations of the RANS case have also been achieved to check the accuracy of the SC and demonstrate its efficiency. In the selected interval about a thousand samples are needed to statistically converge (maximum number

Table 1.1. Variation of the geometrical angle of attack α_w and the inlet flow angle α_e with the random perturbation parameter for 9 samples.

Sample	#1	#2	#3	#4	#5	#6	#7	#8	#9
γ	−1	−0.93	−0.71	−0.38	0	0.38	0.71	0.93	1
α_w	8	8	8	8	8	8	8	8	8
α_e	6	5.86	5.42	4.76	4	3.24	2.58	2.14	2
$\Delta\alpha_e$	2	1.86	1.42	0.76	0	−0.76	−1.42	−1.86	−2

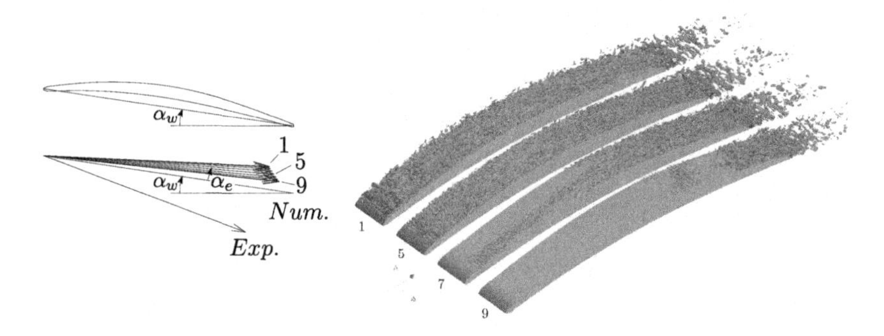

Fig. 1.3. Isosurfaces of the {Q factor ($QC^2/U_0^2 = 2000$) coloured by velocity magnitude for inflow conditions corresponding to LES computations # 1,5,7,9.

of samples $P = 1000$), which makes the stochastic collocation based on Clenshaw-Curtis quadrature method about two orders of magnitude faster for the present study, as it will be illustrated later.

5.3. *UQ results*

5.3.1. *Flow topology*

Changes in flow kinematics with the inlet flow angles α_e are described by iso-surfaces of Q factor in Fig. 1.3. This second invariant of the velocity-gradient tensor helps define vortices and visualize the turbulence development. In LES #1 and #5 (reference), the flow around the airfoil is found similar as in previous studies.[5,6] Small instabilities form close to the reattachment point of the laminar recirculation bubble that trigger transition. The flow tends to re-laminarize toward mid-chord due to the favourable pressure gradient (less turbulent structures and thinner boundary layer). When this gradient becomes adverse at mid-chord, the turbulent boundary layer thickens again and larger turbulent structures appear near the trailing edge. With the increase of the inlet flow angle, more intense structures are

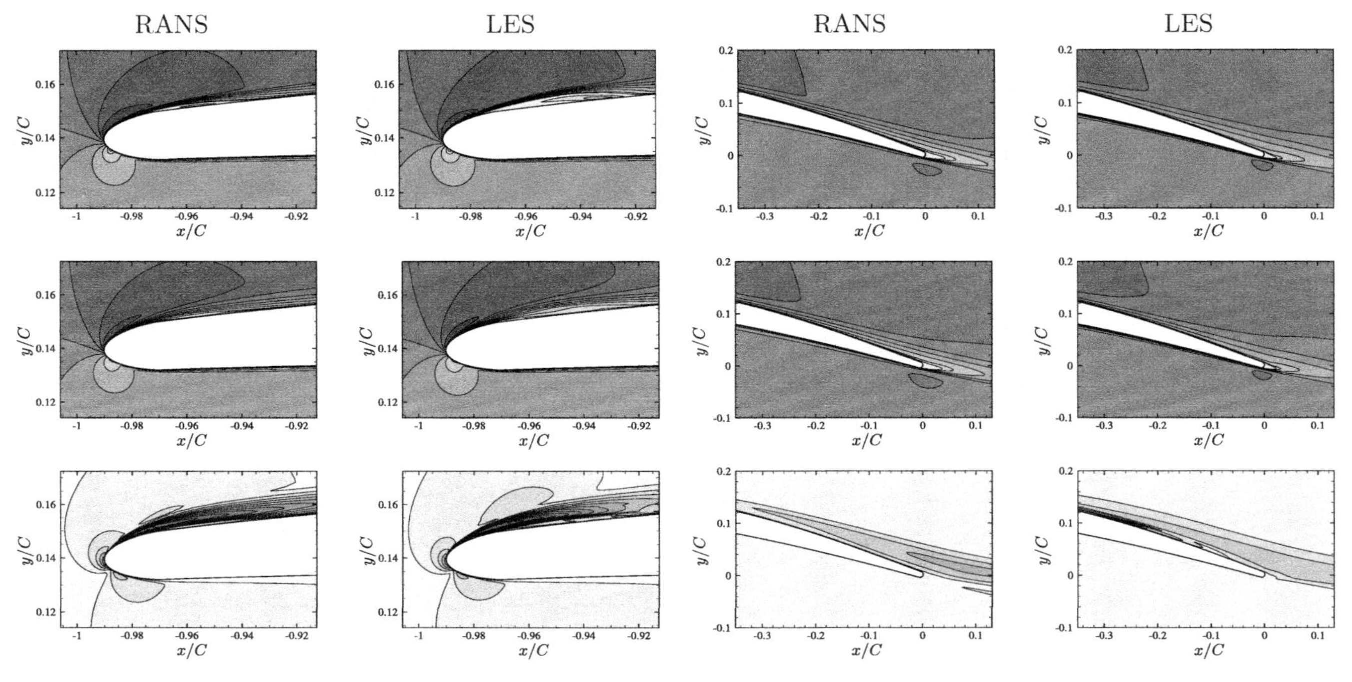

Fig. 1.4. Iso-contours of velocity magnitude around the leading edge and trailing edge. (Top) Deterministic computation # 5, (middle) mean and (bottom) standard deviation of all computations. Contour levels for mean values (0,0.1:0.2:2) and standard deviation (0:0.05:0.75).

formed after the recirculation region at the leading edge; those structures being then convected along the blade chord. In LES #7, the acceleration around the leading edge still yields a weak flow separation at the leading edge, which is not strong enough to trigger transition over the whole span (the Kelvin-Helmholtz instability and roll up has disappeared). Turbulence only develops over a narrow strip and only the adverse pressure gradient after mid-chord triggers transition and the turbulence development over the whole span. In LES #9, the acceleration around the leading edge is no longer strong enough to trigger a flow separation at the leading edge and no transition occurs before mid-chord. Flow separation occurs beyond mid-chord that triggers the transition close to the trailing edge. Instabilities are observed in the recirculation region beyond mid-chord but do not cause any transition in the boundary layer due to their low intensity. Finally, weak vortex shedding is seen in the near-wake on the pressure side for all cases.

Figure 1.4 shows iso-contours of velocity magnitude for the computation #5 (deterministic reference computation) and for the mean and standard deviation from the 9 samples used in the SC method for both RANS and LES computations. These maps are obtained by applying Eq. (1.2) for the time-averaged velocity at each grid point from all the different deterministic computations. Only views around leading and trailing edges are shown since main variations of the flow occur in those regions. For both regions, the flow is found globally similar between the deterministic solution and the mean solution for both RANS and LES computations, and similar between the LES and RANS computations. In the leading-edge region, a larger and thicker recirculation zone is observed in the LES computations, for the deterministic computations #5 and the mean solution. A larger localized production of standard deviation is present in the recirculation region of the RANS computations, caused by a larger variation of the recirculation zone with the inlet flow angle compared with the LES computations. This observation is consistent with the variation of the length and position of the leading-edge recirculation bubble that can be determined from the wall-friction coefficient, as reported in Christophe *et al.*[6] The RANS computations show a monotonic decrease of the size of the recirculation bubble at the leading edge with the decrease of the inlet flow angle from computation #1 to #7, while no systematic decrease of the size of the recirculation bubble is observed in LES computations. In the trailing-edge region, the main difference is appearing beyond mid-chord and before the trailing edge, where a region of higher standard deviation is observed in the LES computations and is not present in the corresponding RANS computations. This

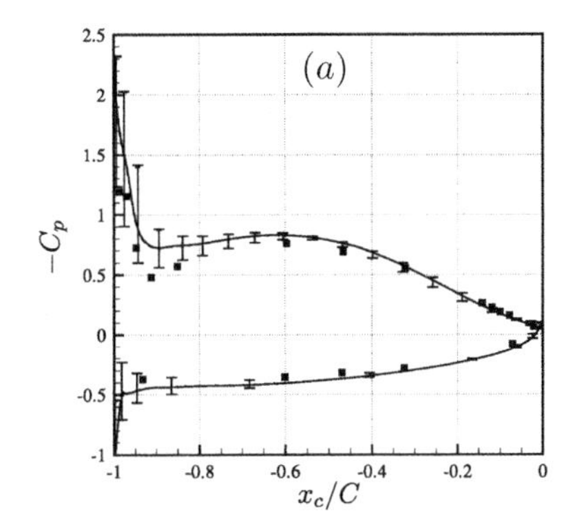

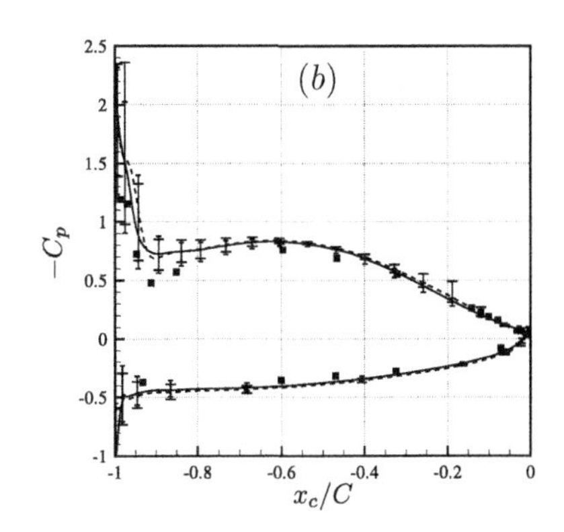

Fig. 1.5. Mean wall-pressure coefficient $-C_p$ (U_0 = 16 m/s, Re_C=1.6×10^5). Experiment[42] (square). (a) Comparison of RANS UQ based methods: SC (plain line and small uncertainty bars) and MC (dashed line and large uncertainty bars). (b) Comparison of flow prediction method using SC: RANS computations (plain line and small uncertainty bars) and LES computations (dashed line and large uncertainty bars).

is caused by the presence of a secondary recirculation zone appearing in the LES computations #8 and #9, and not observed in RANS computations. It should be stressed that this transition in the LES occurs suddenly for a very small variation of incidence in the last two runs (LES #8 and #9).

The differences observed between the RANS and LES computations highlight the complexity of the present problem and the influence of the physical model on the flow topology. The present RANS computations involve fully turbulent flows and therefore cannot correctly take into account laminar and transition regions whereas LES calculations with the present dynamic SGS model do, at least for the location and the early stage of transition.[51]

5.3.2. *Wall-pressure coefficient*

The UQ results on the wall-pressure coefficient $-C_p$ are shown in Fig. 1.5 and compared with experimental results[42] in terms of the mean coefficient (lines) and the intervals, which represent the propagation of 100% of the considered input uncertainty. They are obtained by applying Eq. (1.2) to the wall-pressure distribution. The minimum and maximum of the uncertainty bars are obtained from the minimum and maximum of the response surface from Eq. (1.2). Since the airfoil has a geometrical angle of attack, the chordwise coordinate $x_c = x/\cos\alpha_w$ is used to represent data along the blade chord. Using this coordinate system, the leading edge is located at

$x_c/C = -1$ and the trailing edge is at $x_c/C = 0$. In Fig. 1.5(a), the two stochastic methods are compared using the RANS data set. Both MC and SC methods agree very well yielding the same mean solution and almost identical uncertainty bars. They both show larger uncertainty bars in the leading-edge region and particularly on the suction side in the recirculation bubble. Even though the uncertainty bars are smaller at the trailing edge because of the low value of the mean C_p, the local coefficient of variation is significant. In Fig. 1.5(b), the SC for both RANS and LES are compared. Again, for both RANS and LES computations, larger uncertainty bars are observed in the leading-edge region. Main differences are appearing in the aft region where the second laminar recirculation bubble occurs for the lowest incidences in the LES computations, and not observed in the corresponding RANS computations. If the last two LES results (#8 and #9) were to be removed, the LES would clearly be comparable to the RANS calculations. This shows the complexity of the problem and the large range of flow topologies appearing in the present LES that cannot be captured by the RANS computations due to the simpler RANS model used, as described in the previous section. In regions where the mean simulation result shows a larger difference with respect to the experimental data, the uncertainty bars are also relatively large and account to a large extent for this difference. Both RANS and LES stochastic data sets compare quite well with experiments. The convergence of both stochastic methods for the RANS computations was assessed on this $-C_p$ profiles in Christophe *et al.*[6] The convergence rate of the MC method is found to be $\mathcal{O}(N^{-1})$ while the SC method is found to be $\mathcal{O}(N^{-4})$, which stresses the efficiency of the SC method for a limited number of uncertain variables (curse of dimensionality). Lower convergence rate of the SC method using LES computations is observed because of the large range of flow topologies found in these computations. Similar UQ results for the streamwise boundary-layer velocity profiles, in terms of uncertainty bars and convergence, at the last remote microphone probe (RMP #25, $x_c/C = -0.02$) where the sources of trailing-edge noise are mainly concentrated, can also be found in Christophe *et al.*[6] From these boundary-layer velocity profiles, the necessary boundary-layer parameters for the consequent models of the wall-pressure spectra can be inferred.

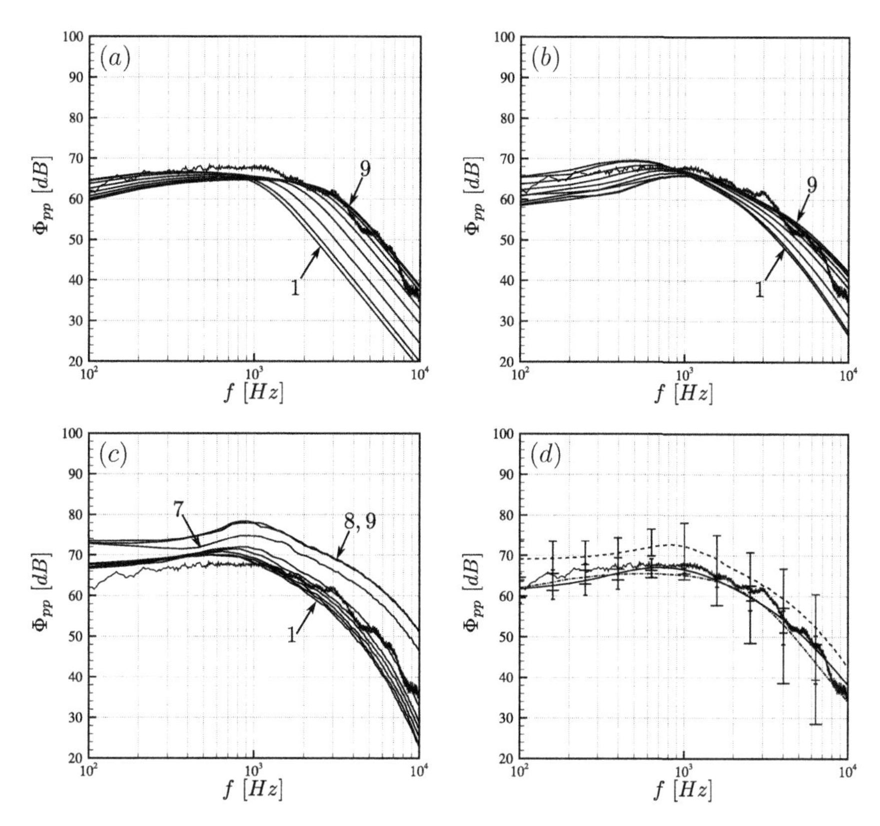

Fig. 1.6. Wall pressure frequency spectra (p_{ref} = 20 mPa) in the trailing-edge area ($x_c/C = -0.02$, RMP # 25) using (a) YR and (b) using PL model for all RANS computations. (c) LES wall-pressure spectra. (d) Comparison of wall-pressure spectra based on RANS and LES inputs in terms of mean and uncertainty bars, YR model (dash-dot and large uncertainty bars), PL model (plain and small uncertainty bars), LES (dash and medium uncertainty bar). Experiment (thin plain).

5.3.3. *Wall-pressure statistics near the trailing edge*

The corresponding trailing-edge spectra of the two methods described above are shown respectively in Figs. 1.6(a) and 1.6(b) for the 9 samples used for SC. A similar behaviour is found for both models where the higher spectrum amplitude at low frequencies corresponds to computation #1 (high inlet flow angle) and the lower one corresponds to computation #9 (lower inlet flow angle). At high frequencies, a reverse behaviour of the models is observed. For both models, a monotonic variation of the wall-pressure spectrum appears going from computation #1 to #9. For both models, a similar crossing is identified around 1 kHz. The YR model presents large variations of the various spectra at high frequencies due to the large uncertainties involved in the wall shear-stress determination, from which the model derives information about the high frequency content. The PL model, not based on the wall shear-stress variable, shows less variations of the wall-pressure spectrum at high frequencies. Yet, at low frequencies, they are larger mainly caused by the slow statistical convergence of the Monte-Carlo integration technique used to integrate the boundary-layer profiles in Eq. (1.11). The corresponding LES spectra for the 9 samples are shown in Fig. 1.6(c) and two groups of LES results are obtained. On the one hand, the LES #1 to #6 are similar as the reference calculation (LES #5) and show similar variations as the RANS reconstructed spectra. On the other hand, the LES #7 to #9 show larger pressure fluctuations for all frequencies due the recirculation bubble close to the trailing edge triggering a more intense turbulent boundary layer than in other cases.

The wall-pressure PSD Φ_{pp} is now used as the stochastic variable in Eq. (1.2). Figure 1.6(d) shows the comparison of all methods with the measured one at RMP #25, in terms of mean and the confidence interval, which represents the propagation of 100% of the considered input uncertainty. Good agreement with experiments is found on the mean wall-pressure spectrum for both methods using RANS information. Larger uncertainty bars are found at high frequencies using the YR model and at low frequencies using the PL model, following explanations given previously. The higher spectra obtained in LES #7 to #9 cause the mean pressure spectrum of the LES to be shifted to higher levels by about 8 dB. Consequently, the LES uncertainty bars are also found to be larger than those of the RANS computations and could be more comparable if the last two simulations were discarded.

5.4. *Stochastic acoustic predictions*

The corresponding sound results are reported in Fig. 1.7, together with the experimental measurements (their uncertainty is ± 1 dB). The far-field acoustic-pressure PSD S_{pp} is the stochastic variable of interest in Eq. (1.2). Mean sound predictions obtained from RANS reconstructed spectra compare favourably with the experimental far-field sound. Similar uncertainty differences between the two RANS-based models YR and PL are observed due to the direct propagation of the uncertainties seen in the trailing-edge wall spectra. Larger uncertainty bars are found at high frequencies using

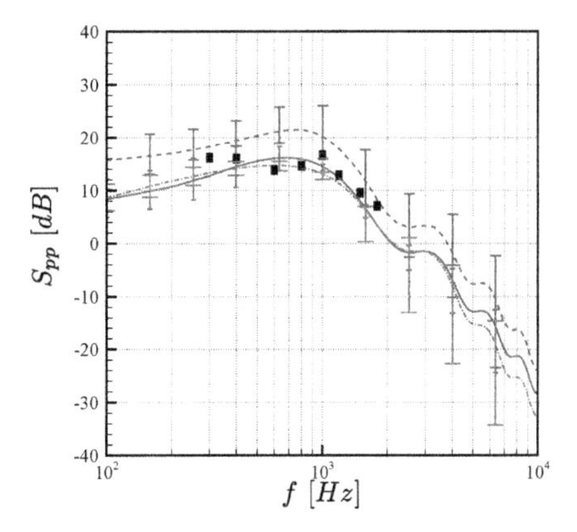

Fig. 1.7. Means and uncertainty bars of far field acoustic spectra (p_{ref}=20 mPa) for the different prediction methods, in the mid span plane above the airfoil ($\theta = 90°$) at $R = 2$ m from the trailing edge. YR model (Dash-dot and large uncertainty bars — red), PL model (plain and small uncertainty bars — blue), LES (dash and medium uncertainty bar — green). Experiment (Square).

the YR model and at low frequencies using the PL models. Again, the LES presents an overestimate compared with the experiments caused by the over-prediction of the wall-pressure spectrum. The corresponding uncertainty bars are then also propagated through the acoustic method.

6. Low-Speed Axial Cooling Fan

6.1. *Deterministic flow simulations*

As mentioned above the low-speed fan considered here is the H380EC1 designed for a best efficiency point at a volume flow-rate $\dot{Q}$ of 2500 m^3/h at a rotational speed Ω of 2500 rpm.[15,38,43,52] The rotor is flush mounted on a plenum walls from which the static pressure rise across the fan Δp is measured. The available experimental data mostly include overall performances and acoustic power measurements at several flow-rates and rotational speeds obtained in a reverberant wind tunnel. The plenum is assumed axisymmetric so that a single blade passage, including the actual tip gap, is modeled and matching nodes are used at the periodic boundary conditions.[52] The flow is modelled with RANS computations using the Shear-Stress-Transport (SST) $k - \omega$ turbulence model in the ANSYS CFX 14 solver. The mesh has a total of 5.1 million hexahedral elements, and is refined in the boundary-layers around the blade to reach dimensionless distances to the wall y^+ less than 5 in the trailing-edge region, where presently most of the sound production occurs. The volume flow-rate is set on the inlet surface of the plenum. The average pressure on the outlet surface is set to the reference pressure. Low-dissipation second-order numerical schemes are used for the flow variables (velocity and pressure), and a first-order scheme for the transport of turbulent quantities. To yield repeatable and consistent results, once the maximum convergence is reached, solutions are averaged over the last 500 iterations.

6.2. *Characterization of the random variables $\vec{\xi}$*

As introduced above, only the parameters defining the operating conditions are considered for uncertainty quantification in a first step. The volume flow rate is defined with a 5% error bound around 2500 m^3/h. The variation of this variable mainly depends on discrete events, such as dirt in the upstream heat exchanger, quality and nature of the fan, and cooling module assembly and manufacturing. A uniform probability density function (PDF) is therefore considered. The variation of rotational speed mainly depends on the motor manufacturing quality and is thus defined by a Gaussian PDF as fit on the manufacturing lines. A usual process dispersion for automotive electrical motors is ± 100 rpm. The rotational speed is thus defined with a 4% error bound around 2500 rpm. For those two variables considered independently, a set of nine RANS computations are determined using a Clenshaw-Curtis quadrature that provides similar convergence as the midspan CD airfoil described in Sec. 5.3.2. For the two-dimensional uncertainty quantification involving both variables, a full tensor grid of 81 RANS computations is used.

6.3. *UQ results*

6.3.1. *Wall-pressure distribution and fan performance*

One dimensional (1D) results on the operating conditions are first analyzed in Fig. 1.8 showing the iso-contours of the standard deviation from the 9 samples used in the SC method for each variable (volume flow rate and rotational speed), on the suction and pressure sides of the blade. The two considered parameters are influencing different zones of the blade surface. For the volume flow rate, a large standard deviation is observed on the suction side of the airfoil, especially at the tip of the blade near the trailing-edge region, due to large variations of the recirculation zone under the

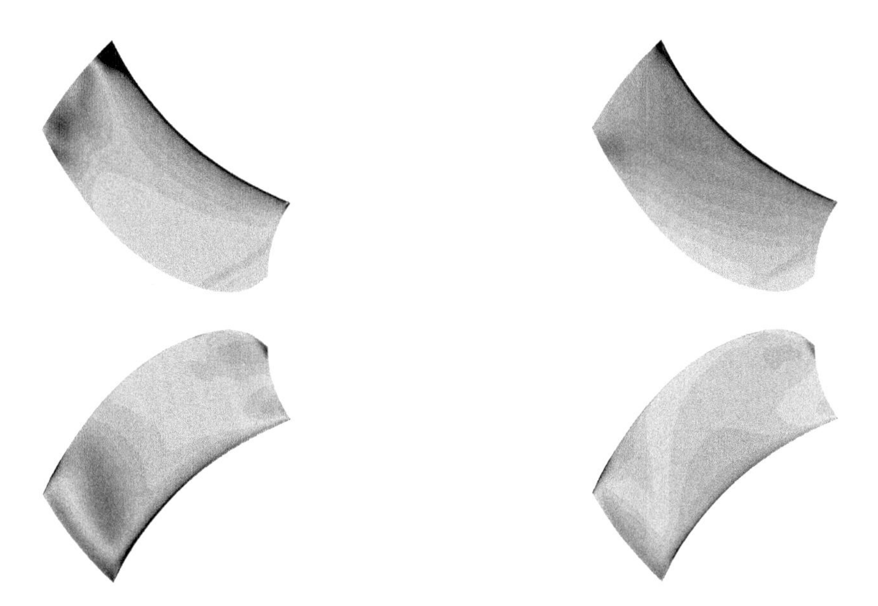

Fig. 1.8. Iso-contours of the standard deviation of the static pressure on (top) suction side (contour levels [0:2:40]) and (bottom) pressure side (contour levels [0:1:20]). (Left) Variation of the volume flow rate and (right) variation of the rotational speed.

rotating ring, as it has been already highlighted in the URANS and SAS simulations on this fan.[52] A second zone of large standard deviation is also observed at the hub on the pressure side corresponding to the flow detachment at the blade leading edge created by the blade cusp. In case of rotational speed variations, the standard deviation of the pressure presents large amplitudes on a broader part of the blade, on both sides, around the leading edge up to mid-chord. The volume flow rate thus mainly influences the size of the recirculation zones while the rotational speed modifies the complete blade pressure distribution.

A comparison of the fan overall performances with the error bar emphasizing the uncertainty interval in case of $\dot{Q}$ variations is presented in Fig. 1.9 for the pressure rise. Two experimental curves are provided as they represent the maximum experimental range of variations obtained on this fan between the many mock-ups and prototypes tested. It should be stressed that the error range on the volume flow rate was chosen to cover the experimental uncertainty, and the resulting uncertainty range on the pressure rise matches the experimental scattering well.

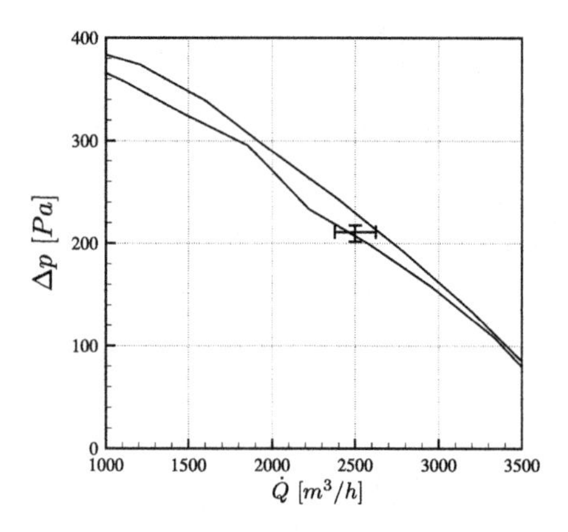

Fig. 1.9. Fan performance.

6.4. *Stochastic acoustic predictions*

In Figs. 1.10–1.12, the spectra of the total simulated sound PoWer Level (PWL) for both YR and PL models are compared with two sets of experimental data, one collected on an engine cooling module in a semi-anechoic chamber[38] and one collected in a reverberant wind tunnel.[53] Both models yield good agreement with the experimental broadband spectra stressing the significant contribution of this noise mechanism at design condition. The PL model presents a small under-prediction of about 3 dB over the whole frequency range, but a better overall shape especially at low frequencies compared with the reverberant wind tunnel data.

Figure 1.10 provides the 1D UQ results for a variation of $\dot{Q}$ for the YR and PL models respectively. The simulated PWL corresponding to the different strips discretizing the blade are shown together with their uncertainty bars. As shown in Moreau *et al.*,[52] the obtained sound spectra have similar shapes, amplitudes and uncertainties than the trailing-edge wall-pressure spectra, showing a direct propagation of the amplitudes and uncertainties through the noise propagation model. Larger uncertainties are obtained in the hub and tip regions as observed previously. The largest uncertainty bars at low frequencies in the tip region correspond to the large standard deviations seen in Fig. 1.8 and are related to the local large-scale flow separation under the fan ring. In the PL model, some lack of convergence in the integration of Eq. (1.11) by a Monte Carlo method could also contribute. At the hub, the largest uncertainties could be traced to the small scale structures created by the blade cusp. The total sound radiated

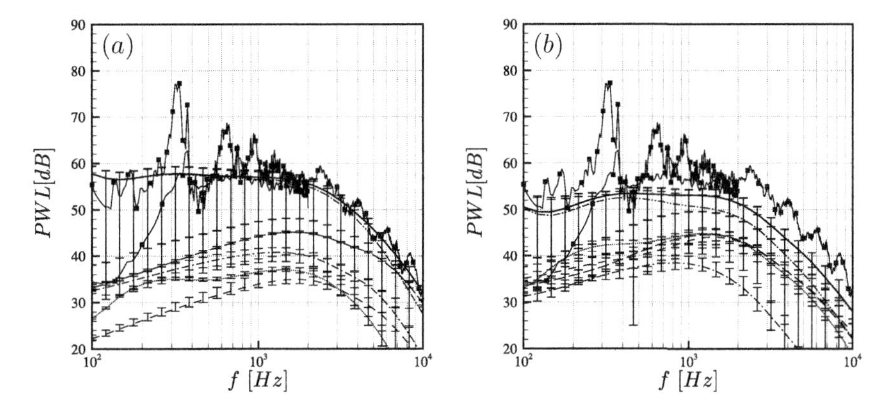

Fig. 1.10. 1D UQ on $\dot{Q}$: (a) YR model and (b) PL model. Total sound (solid bold), strip 1 (dash), strip 2 (dash-dot), strip 3 (dot), strip 4 (long dash), strip 5 (dash-dot-dot), experiment[38,53] (symbols). $p_{ref} = 20$ mPa.

by the complete blade is mainly caused by the noise emitted by the tip strip #5 with almost similar mean amplitudes. At high frequencies, strip #4 also contributes to the total sound spectrum for the YR model while almost all blade strips contribute for the PL model. Moreover, the uncertainty bars for the complete blade are reduced compared with those of the tip strip. The complete fan noise spectrum is thus dominated in amplitude by the tip region while the uncertainty bars are related to uncertainties observed along the complete blade span. A proper control of the flow along the complete blade span is then necessary to reduce the uncertainties on the radiated sound. Finally, the PL model seems to yield lower overall uncertainty than the YR model.

Figure 1.11 shows the corresponding 1D UQ results for a variation of Ω for both YR and PL models. The same mean total spectrum dominated by the tip strip is obtained. The uncertainties (less than 1 dB) are however much smaller than for a variation of flow rate. Therefore the process dispersion on the rotational speed does not trigger significant uncertainties on the fan trailing-edge noise.

Figure 1.12 shows the corresponding 2D UQ results for variations of both $\dot{Q}$ and Ω for the YR and PL models respectively. A similar mean total spectrum dominated by the tip strip is again obtained. The uncertainty bars for both models now lie in between both 1D UQ results on each separate performance parameter, yielding 2 to 6 dB for the YR model and 3 to 4 dB. The PL model therefore gives smaller uncertainties than the YR

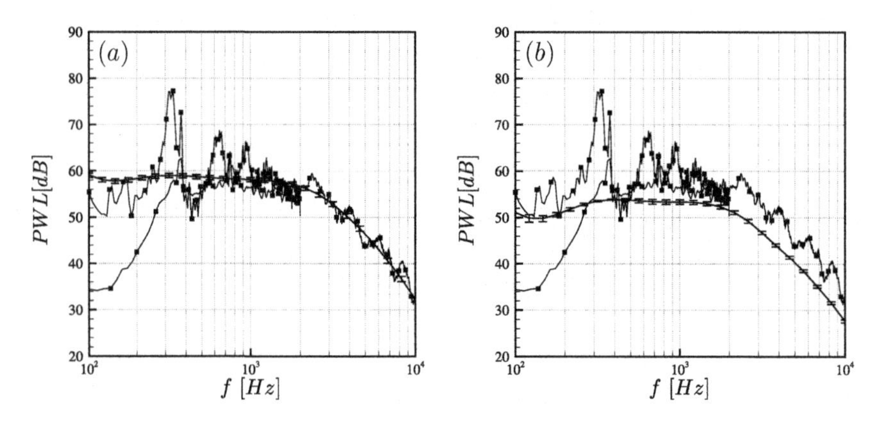

Fig. 1.11. 1D UQ on Ω: (a) YR model and (b) PL model. Total sound (solid bold) and experiment[38,53] (symbols). $p_{ref} = 20$ mPa.

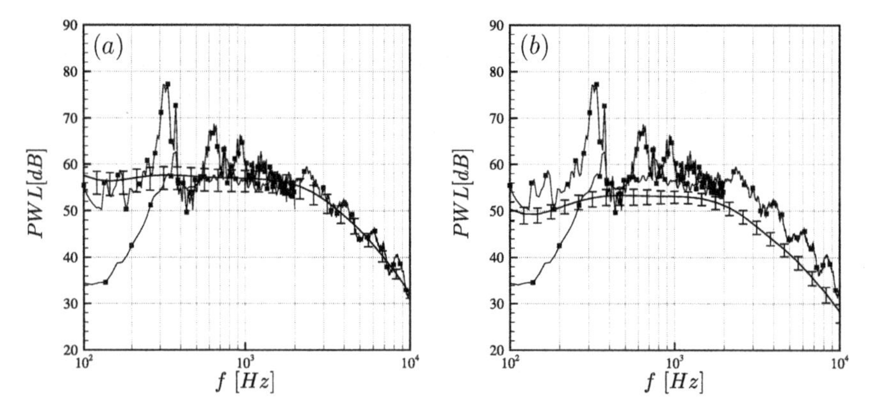

Fig. 1.12. 2D UQ: (a) YR model and (b) PL model. Total sound (solid bold) and experiment (symbols). $p_{ref} = 20$ mPa.

model, more uniformly spread over the whole frequency range.

7. Conclusions

The uncertainty quantification (UQ) related to the self-noise prediction based on a RANS flow computation of a low-subsonic axial fan has been achieved. As the methodology used for fan noise prediction is based on airfoil theories, the uncertainty quantification of a low-speed Controlled-Diffusion (CD) airfoil has been first considered. In both applications, the noise predictions are obtained using Amiet's theory requiring the wall-pressure spectrum near the trailing-edge, provided by two distinct representative models reconstructing the wall-pressure spectra from RANS inputs: the deterministic YR model directly based on the integral parameters of the boundary layer and the PL statistical model based on the velocity field in the boundary layer.

In the airfoil case, the uncertainty is introduced as random velocity components at the inlet of a restricted computational domain, embedded in the jet potential core of the full anechoic wind tunnel. These random variables are intended to model the actual experimental uncertainty in the probe position and in the measurement of the reference velocity, and the numerical inaccuracies in the prediction of the jet development and deflection. To investigate the accuracy of the RANS based method, fully converged unsteady turbulent flow predictions are also provided by incompressible LES. Two deterministic incompressible flow solvers have then been coupled with a non-intrusive stochastic Galerkin method based on a SC to propagate these aerodynamic uncertainties. The flow topology associated with both types of simulations has revealed several noticeable differences. Only in the LES, at the lowest inlet flow angles, the laminar recirculation bubble at the leading edge disappears and a new one forms in the aft of the airfoil. The latter triggers an 8 dB increase in the wall-pressure spectra near the trailing edge. Such a flow bifurcation evidenced by LES could explain the noise increase of similar low-Reynolds number thin airfoils at low inlet flow angles and the noise increase of low-speed fans at high flow rates.[54] As a consequence, the LES-SC has larger uncertainty bars than the RANS-SC in the wall-pressure spectra and consequently in the far field noise over most frequencies. The RANS-UQ does not provide the good trend with incidence and misses the flow bifurcation and the shift of the recirculation bubble to the aft of the profile. When comparing both RANS-SC, both YR and PL models yield very small uncertainty bars around 1 kHz. The YR model has large increasing uncertainty bars at high frequencies caused by its strong dependence on the random wall shear-stress that has a broad probability density function and carries large uncertainties. On the contrary, the PL model has much smaller uncertainty bars at high frequencies but carries more uncertainty at the low frequencies caused by the slow convergence of the Monte Carlo technique used in this model.

In the fan case, realistic 5% and 4% errors about the mean are introduced on the volume flow-rate and the rotational speed respectively to account for the actual experimental and process scattering. The RANS simulations of the fan mounted on a typical industrial test plenum have been run for each parameter yielding a total of 81 calculations. The resulting uncertainty bars obtained on the overall pressure rise match the experimental scattering quite well. By looking at the flow topology and particularly at the wall-pressure field, the variations are mainly located in the hub and the tip region under the fan rotating ring, resulting in large error bars on the corresponding wall-pressure spectra caused by the local flow separations. The final noise spectra of the complete fan are dominated in amplitude by the tip region while the uncertainty bars are related to uncertainties observed along the complete blade span. A proper control of the flow along the complete blade span is then necessary to reduce the uncertainties on the radiated sound. The uncertainties introduced by the variations on volume flow rate are much larger than those given by the rotational speed, and the combined effect of both fan parameters yield a 2-6 dB uncertainty on the far-field noise, with larger variations obtained with the YR model.

The convergence of the SC based on the RANS data set has been verified for both airfoil and fan cases by extending the number of terms in the Curtis-Clenshaw sparse quadrature and by comparing with a Monte Carlo simulation. In the airfoil case, the efficiency of the SC method is clearly demonstrated as it provides the same accuracy as the Monte Carlo approach with a limited number of terms (9) and is two orders of magnitude faster.

Acknowledgements

The authors would like to thank Stanford University and Calcul Quebec (Compute Canada) for providing the necessary computational resources, and C. Hamman and G. Iaccarino for their support during the 2010 and 2012 CTR Summer Programs. This work was supported through the FP7-ECOQUEST project (Grant Agreement no 233541).

References

1. S. Caro and S. Moreau, Aeroacoustic modeling of low pressure axial flow fans. In *6th AIAA/CEAS Aeroacoustics Conference* (2000). AIAA-2000-22094.
2. H. H. Hubbard and K. P. Shepherd, Aeroacoustics of large wind turbines, *J. Acoust. Soc. Am.* **89**(6), 2495–2508 (1991).
3. B. A. Singer, D. P. Lockard and K. S. Brentner, Computational aeroacoustic analysis of slat trailing-edge flow, *AIAA J.* **38**(9), 1558–1564 (2000).

4. S. Moreau, G. Iaccarino, S. Kang, Y. Khalighi and M. Wang, Numerical simulation of a low speed fan blade. In *Proceedings of the Summer Program 2004*, Centre for Turbulence Research, Stanford Univ./NASA Ames (2004).
5. S. Moreau, D. Neal, Y. Khalighi, M. Wang and G. Iaccarino, Validation of unstructured-mesh les of the trailing-edge flow and noise of a controlled-diffusion airfoil. In *Proceedings of the Summer Program 2006*, Centre for Turbulence Research, Stanford Univ./NASA Ames (2006).
6. J. Christophe, S. Moreau, C. Hamman, J. Witteveen and G. Iaccarino, Uncertainty quantification for the trailing-edge noise of a controlled-diffusion airfoil, *AIAA J.* **53**(1), 42–54 (2015).
7. Y. Addad, R. Prosser, D. Laurence, S. Moreau and F. Mendonca, On the use of embedded meshes in the les of external flow, *Flow Turbulence Combust.* **80**, 392–403 (2008).
8. M. Wang, S. Moreau and M. Roger, LES prediction of wall-pressure fluctuations and noise of a low-speed airfoil, *Internat. J. Aeroacoustics* **8**(3), 177–198 (2009).
9. J. Christophe, J. Anthoine and S. Moreau, Trailing edge noise of a controlled-diffusion airfoil at moderate and high angle of attack. In *15th AIAA/CEAS Aeroacoustics Conference* (2009). AIAA-2009-3196.
10. S. Moreau, M. Roger and J. Christophe, Flow features and self-noise of airfoils near stall or in stall. In *15th AIAA/CEAS Aeroacoustics Conference* (2009). AIAA-2009-3198.
11. J. Winkler, S. Moreau and T. Carolus, Large-eddy simulation and trailing-edge noise prediction of an airfoil with boundary-layer tripping. In *15th AIAA/CEAS Aeroacoustics Conference* (2009). AIAA-2009-3197.
12. L. Corriveau, S. Moreau, M. Roger and J. Christophe, Experimental and numerical unsteady acoustic sources and self-noise of a katana blade. In *16th AIAA/CEAS Aeroacoustics Conference* (2010). AIAA-2010-3804.
13. J. Winkler, T. Carolus and S. Moreau, Airfoil trailing-edge blowing: broadband noise prediction from large-eddy simulation, *AIAA J.* **50**(2), 1558–1564 (2012).
14. M. Sanjosé, S. Moreau, M. Kim and F. Pérot, Direct self-noise simulation of the installed controlled diffusion airfoil. In *17th AIAA/CEAS Aeroacoustics Conference* (2011). AIAA-2011-2716.
15. D. Casalino, S. Moreau and M. Roger, One, no one and one hundred thousand methods for low-speed fan noise prediction, *Internat. J. Aeroacoustics* **3**(3), 307–327 (2010).
16. S. Glegg, B. Morin, O. Atassi and R. Reba, Using Reynolds-averaged Navier–Stokes calculations to predict trailing edge noise, *AIAA J.* **48**(7), 1290–1301 (2010).
17. Y. Rozenberg, S. Moreau, M. Henner and S. C. Morris, Fan trailing-edge noise prediction using RANS simulations. In *16th AIAA/CEAS Aeroacoustics Conference* (2010). AIAA-2010-3720.
18. S. Remmler, J. Christophe, J. Anthoine and S. Moreau, Computation of wall-pressure spectra from steady flow data for noise prediction, *AIAA J.* **48**(9), 1997–2007 (2010).

19. J. Christophe, S. Moreau, C. Hamman, J. Witteveen and G. Iaccarino, Uncertainty quantification for the trailing-edge noise of a controlled-diffusion airfoil. In *Proceedings of the Summer Program 2010* (2010).
20. J. Christophe and S. Moreau, Uncertainty quantification of low-speed fan noise. In *Proceedings of the Summer Program 2012*, Centre for Turbulence Research, Stanford Univ./NASA Ames (2012).
21. W. Oberkampf, T. Trucano and C. Hirsch, Verification, validation and predictive capability in computational engineering and physics, *Appl. Mech. Rev.* **57**(5), 345–384 (2004).
22. R. H. Schlinker and R. K. Amiet, Helicopter rotor trailing edge noise. Technical Report. NASA CR 3470 (1981).
23. S. Moreau, M. Henner, G. Iaccarino, M. Wang and M. Roger, Analysis of flow conditions in freejet experiments for studying airfoil self-noise, *AIAA J.* **41**(10), 1895–1905 (2003).
24. R. L. Panton and J. H. Linebarger, Wall pressure spectra calculations for equilibrium boundary layers, *J. Fluid Mech.* **65**(02), 261–287 (1974).
25. J. Christophe, M. Sanjosé and S. Moreau, Uncertainty quantification of a low-speed axial fan self-noise. In *Proceedings ISROMAC Conference 2012* (2012).
26. M. Roger and S. Moreau, Back-scattering correction and further extensions of amiet's trailing-edge noise model. Part 1: theory, *J. Sound Vib.* **286**, 477–506 (2005).
27. R. Walters and L. Huyse, Uncertainty analysis for fluid mechanics with applications. Technical Report NASA CR-2002-211449 (2002).
28. D. Cacusi, *Sensitivity and Uncertainty Analysis.* CRC Press (2003).
29. D. Lucor, D. Xiu, C. Su and G. Karniadakis, Predictability and uncertainty in CFD, *Int. J. Numer. Methods Fluids* **43**(5), 483–505 (2003).
30. G. Iaccarino, Quantification of uncertainty in flow simulations using probabilistic methods, *VKI Lecture Series — Nonequilibrium Gas Dynamics* (2008).
31. H. N. Najm, Uncertainty quantification and polynomial chaos techniques in computational fluid dynamics, *Ann. Rev. Fluid Mech.* **41**, 35–52 (2009).
32. M. A. Tatang, W. Pan, R. G. Prinn and G. J. McRae, An efficient method for parametric uncertainty analysis of numerical geophysical models, *J. Geophys. Res.* **102**(D18), 21,925–21,922 (1997).
33. I. Babuska, F. Nobile and R. A. Tempone, A stochastic collocation method for elliptic partial differential equations with random input data, *SIAM J. Numer. Anal.* **45**(3), 1005–1034 (2007).
34. J. C. Chassaing and D. Lucor, Stochastic investigation of flows about airfoils at transonic speeds, *AIAA J.* **48**(5), 938–950 (2010).
35. C. W. Clenshaw and A. R. Curtis, A method for numerical integration on an automatic computer, *Numer. Math.* **2**, 197–205 (1960).
36. R. K. Amiet, Noise due to turbulent flow past a trailing edge, *J. Sound Vibration.* **47**(3), 387–393 (1976).
37. M. J. Lighthill, On sound generated aerodynamically. I. general theory, *Proc. R. Soc. L. Series A.* **211**(1102), 564–587 (1952).

38. S. Moreau and M. Roger, Competing broadband noise mechanisms in low-speed axial fans, *AIAA J.* **45**(1), 48–57 (2007).
39. S. Sinayoko, M. Kingan and A. Agarwal, Trailing edge noise theory for rotating blades in uniform flow, *Proc. R. Soc. A* **469**, 20130065 (2013).
40. D. Coles, The law of the wake in the turbulent boundary layers, *J. Fluid Mech.* **1**(2), 191–226 (1956).
41. Y. Rozenberg, *Modélisation analytique du bruit aérodynamique à large bande des machines tournantes: utilisation de calculs moyennés de mécanique des fluides,* PhD thesis, Ecole Centrale de Lyon (2007).
42. S. Moreau and M. Roger, Effect of airfoil aerodynamic loading on trailing-edge noise sources, *AIAA J.* **43**(1), 41–52 (2005).
43. S. Magne, M. Sanjosé, S. Moreau and A. Berry, Aeroacoustic prediction of the tonal noise radiated by a ring fan in uniform inlet flow. In *18th AIAA/CEAS Aeroacoustics Conference* (2012). AIAA-2012-2122.
44. F. Menter, Two-equation eddy viscosity turbulence models for engineering applications, *AIAA J.* **32**(8), 1598–1605 (1994).
45. M. Germano, U. Piomelli, P. Moin and W. H. Cabot, A dynamic subgrid-scale eddy viscosity model, *Phys. Fluids A* **3**(7), 1760–1765 (1991).
46. D. Lilly, A Proposed modification of the Germano Subgrid-Model Closure Method, *Phys. Fluids A* **4**(3), 633–635 (1992).
47. J. Kim and P. Moin, Application of a fractional-step method to incompressible Navier–Stokes equations, *J. Comput. Phys.* **59**, 308–323 (1985).
48. K. Mahesh, G. Constantinescu and P. Moin, A numerical method for large-eddy simulation in complex geometries, *J. Comput. Phys.* **197**, 215–240 (2004).
49. F. Ham and G. Iaccarino, Energy conservation in collocated discretization schemes on unstructured meshes. In *Annual Research Briefs*, Centre for Turbulence Research, Stanford Univ./NASA Ames (2004).
50. P. Sagaut, *Large Eddy Simulation for Incompressible Flows.* Springer-Verlag (2002).
51. T. Sayadi and P. Moin, Predicting natural transition using large eddy simulation. In *Annual Research Briefs*, Centre for Turbulence Research, Stanford Univ./NASA Ames (2011).
52. S. Moreau, M. Sanjosé, S. Magne and M. Henner, Aerocoustic predictions of a low-subsonic axial fan. In *Proceedings ISROMAC Conference 2012* (2012).
53. M. Sanjosé, D. Lallier-Daniels and S. Moreau, Aeroacoustics analysis of a low subsonic axial fan. In *ASME Turbo Expo 2015 Conference*, number ASME GT2015-12345, Montreal, Canada (June 15–19, 2015).
54. R. E. Longhouse, Noise separation and design considerations for low tip-speed and axial-flow fans, *J. Sound Vibration.* **4**(48), 461–474 (1976).

Chapter 2

Computational Simulations of Near-Ground Sound Propagation: Uncertainty Quantification and Sensitivity Analysis

Chris L. Pettit

Aerospace Engineering Department,
United States Naval Academy
590 Holloway Road, Mail Stop 11-B
Annapolis, MD 21402, USA
pettitcl@usna.edu

D. Keith Wilson

Signature Physics Branch,
U.S. Army Engineer Research and Development Center
Cold Regions Research and Engineering Laboratory
72 Lyme Road, Hanover, NH 03755, USA
D.Keith.Wilson@usace.army.mil

We review the state of the art in uncertainty quantification (UQ) and sensitivity analysis (SA) of near-ground sound propagation (NGSP) based on computational models. This review is unique in being organized around the core concept of a computational mechanics error budget. We outline phenomena essential to simulating near-ground sound, and how errors and uncertainties in the common models fall into the error budget. This description forms a basis for examining the connections between the error budget and recent UQ and SA research in NGSP simulations and measurements. Given the myriad forms of uncertainty in NGSP simulations and the impossibility of reducing all of them to negligible levels, we claim that the error budget can provide insight into the relative benefits of future research directions in NGSP simulations.

1. Introduction

Near-ground sound propagation (NGSP) is the branch of atmospheric acoustics in which the presence of the ground is central to determining how acoustic energy is transmitted through the atmosphere.

Many researchers, e.g., Dosso,[1] Finette[2,3] and Pettit and Wilson,[4] have highlighted the need to account thoroughly for uncertainty in computational predictions of outdoor acoustics and underwater acoustics. Valente *et al.*[5] described "... high variability in the characteristics of blast sound propagation at distances up to 16 km." Their measurements suggest that "... instantaneous meteorological conditions are much more important predictors of sound propagation than location, terrain, and climate."

Our specific goals in this chapter are to

- Objectively summarize uncertainty quantification (UQ) and sensitivity analysis (SA) methods used in computational simulations of NGSP, with an error budget[6] (see Sec. 2.1) as a unifying theme.
- Highlight how the chosen physical models and sampling-based UQ methods affect the physical interpretations and statistical convergence rate; e.g., see Ref. 7.

We begin by describing phenomena essential to simulating NGSP. This description forms a basis for considering the suitability of computational physics models of NGSP as well as methods for quantifying the sensitivities and uncertainties in their results. In Sec. 2 we describe Ghanem's[6] version of a general error budget and outline theoretical and practical aspects of computational models of NGSP. The presentation of the theory is cursory but deep enough for non-acousticians to benefit from what follows. We believe the presentation in Sec. 2 is unique in attempting to arrange NGSP modeling and measurement concerns under the error budget rubric. Along the way, we cite studies in which one or more components of the error budget were prominent. In most cases the error budget we describe is implicit in the cited work. We seek to make the connections clear. Section 3 highlights the importance of UQ and SA as complementary ways to characterize the error budget in NGSP applications. Section 4 builds on Sec. 3 by citing recent studies that may be related to particular components of the computational NGSP error budget. Section 5 contains our conclusions.

1.1. *A brief overview of near-ground sound propagation phenomena*

By *near ground*, we mean the atmospheric surface layer (ASL). The ASL is the part of the atmosphere in which the momentum stress and heat flux differ by less than 10% from their surface values. It typically extends to an altitude of 50 m to 200 m above the ground.[8] Sound propagation is influenced by atmospheric structure at altitudes up to about 10% of the horizontal distance between the source and receiver, so distances may extend up to 2 km. Our focus is on propagation in the ASL, but many of the principles apply to propagation in the atmosphere more generally. Focusing on the ASL facilitates a more consistent modeling approach with well-defined parametric representations.

NGSP is affected by many complex, interacting phenomena, as illustrated in Fig. 2.1. A full description of the sound propagation problem requires quantifying the acoustic field properties of the atmosphere, ground, vegetation, and man-made structures, as well as the shape of the natural topography. The problem is too complex to be thoroughly described here, but we describe enough to support the rest of the chapter. More complete descriptions are readily available in the specialized literature.[9–12]

Consider for simplicity a point source. Sound energy from the source in an infinite, homogeneous, quiescent atmosphere would spread spherically while being absorbed through viscous effects and molecular relaxation, which depend on humidity and temperature (see App. B in Ref. 11). Near the ground, however, the sound is refracted through vertical gradients of temperature and wind velocity. The refraction effects are very important in determining the loudness of sound on the ground. Sound is also scattered by turbulence and other random atmospheric variations such as internal gravity waves.

Mean refraction in the ASL may be upward or downward, depending on whether the propagation is upwind or downwind, and whether the temperature gradient is negative or positive, respectively. The mean wind speed profile in the direction of propagation and the mean temperature profile combine to form an effective sound speed profile, which often varies importantly in space and time. Upwind propagation commonly causes near-ground sound rays to be refracted upward, thereby producing near-ground refractive shadow zones into which the sound energy is propagated through scattering by turbulence. Downwind propagation generally leads to downward refraction of higher-angle sound rays and ducting of waves, which increases the concentration of acoustic energy near the ground, where ground reflections may then lead to acoustic interference regions.

Sound waves are also partially reflected and absorbed by the ground surface. The ground's effects on the incoming sound are quantified in terms of the specific acoustic impedance, which is defined as the ratio of acoustic pressure to particle velocity at the surface. Impedance models put forth for simulating NGSP[11–13] commonly assume the ground's impedance is constant along the propagation path. Reflected sound waves have different

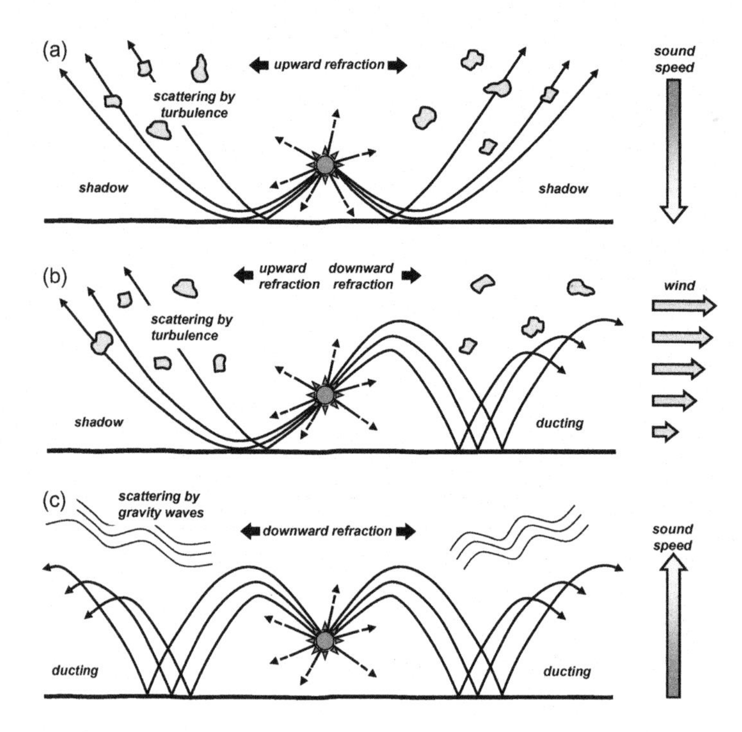

Fig. 2.1. Near-ground sound propagation (NGSP) phenomena. (a) Refraction with a negative temperature (sound speed) gradient and low wind, as characteristic of sunny, fair weather. (b) Refraction with a strong wind and a negligible temperature gradient, as characteristic of cloudy weather. (c) Refraction with a positive temperature (sound speed) gradient and low wind, as characteristic of clear night, fair weather conditions.

magnitude and phase from the incoming waves, and subsequently undergo interference with other sound waves near the ground.

1.2. *Motivation for this chapter*

High-fidelity computational mechanics models are available for predicting NGSP. However, the accuracy of predictions is constrained by many uncertainties and practical limitations. This chapter summarizes recent work in assessing the main sources of uncertainty and how they govern the resulting uncertainty in computational models of NGSP.

We view uncertainty and error in high-fidelity NGSP simulations as a combined state of imperfect knowledge. This requires viewing computational models of physical systems as a separate system composed of $\{physics\ models\} + \{data = measurements + statistics\} + \{numerics\}$. Each of these components may be compromised by uncertainty and error. In principle, each component can be parametrized, such that the uncertainties and errors are encoded in these generalized parameters. At that point, SA could be invoked to see which parameters strongly influence the predictions and thereby define the model's *error budget*.[6] While we are not so bold as to attempt such a far-reaching parametrization of the error budget here, we will use this concept as scaffolding for much of the discussion that follows.

In NGSP high-fidelity models, imperfect knowledge may be found throughout a given application, such as inadequate resolution of the terrain's local topography and the atmosphere's and ground's acoustic properties, random environmental variations such as turbulence, and the practical difficulty of accounting for periodic variations and long-term trends in vegetation and land use. Medium-fidelity methods, such as the parabolic equation method described in Sec. 2.2, are further compromised by uncertainty about the severity of their basic assumptions in a given application, and whether these assumptions change the nature of the uncertainty in generic model outputs, such as relative sound pressure level (SPL) throughout an area interest; for example, NGSP predictions often become increasingly untrustworthy as they approach the validity boundaries of key assumptions, but assumptions may be more or less benign depending on the range of source frequencies to be studied, the prevalence of upwind or downwind propagation, and many other application-dependent factors that may confound efforts to assess the validity of predicted sound levels.

The severity of these practical limitations on predictive skill depends also on how their results will be used. This concern is common in all applications of computational mechanics, but the practical and regulatory implications of inaccurate predictions are beyond our scope here. Instead, we restrict the discussion to the this thesis: the credibility and usefulness of predictions of near-ground sound propagation under uncertainty will be promoted by examining what Ghanem[6] called a model's *error budget.*

The error budget is an epistemic construct, so it cannot be uniquely defined or computed exactly. However, an agreed-upon form can be described in a thorough, self-consistent way. We use Ghanem's decomposition of the budget into these contributors: incorrect physics and spatio-temporal resolution, limited data for characterizing the propagation environment, inaccuracies in the uncertainty models (e.g., assuming a distribution family unsuited to the data or the underlying physics), and the potential for interactions between limited data and the fitting of probabilistic models.

UQ and SA offer methods for understanding the properties of the error budget and for identifying missing data that could reduce it. We focus on sampling-based methods, e.g., Monte Carlo simulation (MCS) and its many variants, because they are straightforward to implement and understand. In particular, they enable what is sometimes referred to as *non-intrusive UQ*,[14,15] i.e., UQ in which the computational algorithm need not be redesigned and rewritten before UQ can be done. Sampling is computationally costly in realistic applications, so we consider methods to make good use of each sample and to replace direct sampling with sampling and calculus-based analysis of faster surrogate models based on the samples.

As noted above, many researchers have called for better management of uncertainty in computational predictions of outdoor and underwater acoustics. Their observations motivate a comprehensive approach to quantifying and parsing the NGSP error budget. Computational models should be used to predict variability across the range of expected operating conditions, sensitivity analysis should be used to assess the importance of the model's inputs in driving the predicted variability, and experiments should be used to evaluate the prediction methods in realistic benchmark settings.

2. Central Topics in Computational Modeling of Near-Ground Sound Propagation Under Uncertainty

2.1. *Error budgets in computational NGSP as a predictive science*

Ghanem[6] decomposed a model's error budget between the true value of a stochastic response U and its predicted value $\widehat{U}$ as follows:

$$\widehat{U} - U = \varepsilon_m + \varepsilon_{d|m} + \varepsilon_{p|d,m} + \varepsilon_{h|p,d,m}.$$

- $U : (\Omega, \mathcal{S}, P) \to \mathbb{R}^N \times \mathbb{R}$ is the correct stochastic response process, possibly as a function of both space and time, hence the mapping from the probability space $(\Omega, \mathcal{S}, P)$ to $\mathbb{R}^N \times \mathbb{R}$.
- ε_m — which can be reduced through better physics models.
- $\varepsilon_{d|m}$ — which can be reduced through better data, e.g., more accurate and precise measurements.
- $\varepsilon_{p|d,m}$ — which can be reduced through better statistics and probability models.
- $\varepsilon_{h|p,d,m}$ — which can be reduced through better numerical methods and grids.

In our review of the literature, we could not find specific names for these error budget components.

In computational NGSP, the parabolic equation (PE) model is a common reference model. See Sec. 2.2.2 for details of the PE model. The PE model must be coupled with models of the ground impedance and the sound speed variation in the ASL to completely specify an NGSP simulation. Once these models are chosen, ε_m is fixed, but the other three error terms remain undetermined.

Leroy *et al.*[16] described the importance of estimating what they called *combined uncertainty*, i.e., "...a combination of model uncertainty, measurement uncertainty, and physical variability." This appears to be the closest example in the NGSP literature of an error budget description similar to Ghanem's. Leroy *et al.* almost certainly meant model uncertainty and measurement uncertainty to be the same as the first two terms in Ghanem's error budget, ε_m and $\varepsilon_{d|m}$. What they meant by uncertainty related to physical variability is less clear. They might have meant the spatio-temporal variability in the environment (i.e., aleatory uncertainty), but this misses the fact that how physical variability is represented in simulations also injects epistemic uncertainty through the analyst's selection of probability models. The error budget's third term, $\varepsilon_{p|d,m}$, reflects this intermingling of aleatory and epistemic uncertainties.

Finally, Leroy *et al.* did not include an explicit term in their combined uncertainty that corresponded to $\varepsilon_{h|p,d,m}$. This omission might be justified if numerics and grids were implicit in their model uncertainty term; alternatively, perhaps they took the view that numerical aspects such as spatial discretization should be addressed through grid convergence studies before the uncertainty in the environment is addressed.

We prefer Ghanem's explicit inclusion of this term. It is clearer, more general, and it reveals the potential interactions between the spatial uncertainty in a system's parameters and the spatial resolution of the physics model. This last point is somewhat subtle, easily overlooked, and not explored sufficiently in the computational mechanics UQ literature. While studying the application of stochastic expansions (see Sec. 3.1.2) to quantify uncertainty in geometrically nonlinear plate bending, Pettit and Beran[17] uncovered a "...mild but noticeable sensitivity of solution statistics to spatial discretization that is above that of the deterministic baseline problem... A connection exists between the grid density and accuracy of the expansion coefficients, *even for grids that might be considered converged for modeling typical realizations*" [emphasis added].

The error budget terms may be sensitive to one or more model parameters. These sensitivities ought to be studied to promote efficient use of modeling and data-gathering resources. Wilson *et al.*[7] concluded: "Predictive uncertainties should also be considered when evaluating trade-offs between accuracy and computation time, with regard to both the number of calculations and the complexity of the propagation model. Computational effort should not be needlessly expended in attempting to predict details of the propagation that are, in essence, unpredictable."

2.2. *Physical fidelity of computational NGSP prediction*

2.2.1. *Fundamentals of NGSP prediction*

For brevity and specificity, we restrict our discussion to linear acoustics models, in which fluctuations due to the sound waves are small relative to the ambient (background, or average) values in the medium through which the waves propagate.[18] Nonlinear effects may be neglected unless the sound waves are very loud, such as those due to an explosion or shock. In the linear approximation, the atmospheric fields of pressure, density, and particle velocity are separated into average and fluctuating components, i.e.,

$$p_{\mathrm{a}} = p_{\mathrm{av}} + p, \qquad \rho_{\mathrm{a}} = \rho_{\mathrm{av}} + \rho, \qquad \vec{v}_{\mathrm{a}} = \vec{v}_{\mathrm{av}} + \vec{v},$$

where p_{a}, ρ_{a}, and $\vec{v}_{\mathrm{a}}$ are net values; p_{av}, ρ_{av}, and $\vec{v}_{\mathrm{av}}$ are average values associated with the background atmospheric state; and p, ρ, and $\vec{v}$ are small fluctuations induced by acoustic waves. For acoustical modeling, our primary interest lies in determining p, the sound (or acoustic) pressure, since microphones measure this quantity. In the linear approximation of adiabatic propagation in a motionless, calorically perfect gas ($p_{\mathrm{av}} = R\rho_{\mathrm{av}}T_{\mathrm{av}}$), the equations of mass conservation, linear momentum, and equation of state of the sound field reduce to a 3-D wave equation,

$$\nabla^2 p = \frac{1}{c^2}\frac{\partial^2 p}{\partial t^2}, \tag{2.1}$$

where

$$c = \sqrt{\gamma \frac{p_{\mathrm{av}}}{\rho_{\mathrm{av}}}} = \sqrt{\gamma R T_{\mathrm{av}}} \tag{2.2}$$

is the adiabatic sound speed and γ is the ratio of specific heats.

If the source is harmonic with frequency f, then the acoustic wave will propagate harmonically at this same frequency. At a given location, the received sound waves will generally have a different amplitude and phase than at the source, as dependent upon the propagation distance, attenuation by the atmosphere, ground reflections, and other effects. Customarily, the sound pressure and velocity fields are represented as complex values,

$$p = \mathrm{Re}\left(p_{\mathrm{c}} e^{-i\omega t}\right), \tag{2.3}$$

$$\vec{v} = \mathrm{Re}\left(\vec{v}_{\mathrm{c}} e^{-i\omega t}\right), \tag{2.4}$$

where p_{c} and $\vec{v}_{\mathrm{c}}$ are complex amplitudes $\omega = 2\pi f$ is the angular frequency. Harmonic sound waves therefore depend on time only through the $e^{-i\omega t}$ factor, so the complex pressure amplitude satisfies the Helmholtz equation,

$$\nabla^2 p_{\mathrm{c}} + k^2 p_{\mathrm{c}} = 0, \tag{2.5}$$

where $k^2 = \omega/c = 2\pi/\lambda$ is the wave number and $\lambda = c/f$ is the wavelength.

When sound propagates through a moving medium (e.g., a windy atmosphere), the wave and Helmholtz equations are complicated by the presence of additional terms. This topic is treated thoroughly elsewhere.[19] When the sound propagates at angles within about 20° of horizontal, however, modeling can be greatly simplified through the use of the effective sound-speed approximation. The actual sound speed is replaced with $c_{\mathrm{eff}}(\alpha) = c + v_{\perp} \cos\alpha$, where $v_{\perp}$ is the horizontal wind speed and α the angle between the wind and the azimuthal propagation direction. In general, modeling of sound propagation in the atmosphere depends upon the spatially and temporally varying fields c, ρ_{av}, and $\vec{v}_{\mathrm{av}}$.[19]

2.2.2. *Computational modeling of sound waves in the atmosphere*

Recent advances in computational acoustics,[11,12] including the parabolic equation (PE),[20,21] fast field program (FFP),[22,23] finite-difference time-domain (FDTD) method,[24,25] pseudo-spectral time-domain (PSTD) method,[26] and boundary-element method (BEM),[27] allow many factors impacting NGSP to be addressed realistically. Predictive methods can be broadly distinguished into four categories: (1) heuristic (or engineering) methods, (2) geometric acoustics (ray tracing and related high-frequency approximations), (3) wave-based, frequency-domain methods, and (4) wave-based, time-domain methods. Selecting one depends greatly upon the application and goals of the analysis. FDTD methods are the most versatile, as they support time-varying source characteristics and environmental conditions, atmospheric refraction and turbulent scattering, and reflections from objects, e.g., buildings, barriers, and trees. However, more restrictive methods remain popular because they are less faster and easier. Many engineering scenarios, for example, are too complex for rigorous, time-consuming calculation methods. Heuristic approaches are often adopted to speed up and standardize the predictions. In research settings, more intensive and accurate numerical calculations may be viable.

Here we consider only the non-heuristic, wave-based methods. By *wave-based*, we mean methods that solve the wave, Helmholtz, or related equations for wave propagation in a moving medium. Unlike ray tracing, the computational cost increases with frequency (decreasing wavelength) because the solution is calculated on a spatial grid having resolution proportional to the wavelength. Wave-based solutions may be implemented in the frequency or time domain. The PE and FFP are generally implemented in the frequency domain. Frequency-domain solutions can be calculated one frequency at a time, so less memory is often required. For the time-domain methods, resolution of the temporal grid varies inversely with the maximum supported frequency. Wave-based methods can handle the ground effect either by simulating the wave in the ground directly, or by simulating only the airborne sound and applying an impedance boundary condition.

Direct solution of the wave (or Helmholtz) equation in 3-D is often computationally prohibitive. For reasonable accuracy, space must be discretized much finer than the shortest wavelength of interest; a typical value is $\lambda/10$. Since the wavelength is about 3.3 m at 100 Hz, and 0.33 m at 1000 Hz (a typical frequency range), a 3-D grid with dimensions of $1000\,\mathrm{m} \times 1000\,\mathrm{m} \times 300\,\mathrm{m}$ would require approximately 10^{10} elements at 100 Hz and 10^{13} elements at 1000 Hz. For this reason, fully 3-D solution of Eq. (2.1), e.g. with FDTD methods, is usually attempted only at low frequencies, or 2-D approximate solution methods are employed. The latter can be viewed as assuming azimuthal independence of the propagation when the solution is calculated.[11] This is often called the $N \times 2\mathrm{D}$ method, since N independent 2-D solutions in vertical planes are calculated. In effect, sound energy propagating in directions normal to each vertical plane is neglected.

For further computational efficiency, one-way spatial marching algorithms are often employed to calculate the propagation within each vertical plane. The solution is marched radially outward from the source, similarly to time-marching when solving the diffusion equation. Such one-way algorithms originate from the replacement of the Helmholtz equation (2.5), which is elliptic, with a parabolic approximation. The derivation of the parabolic equation (PE) model is too lengthy to provide here; see App. G in Ref. 11 for details. In what follows, terms with subscript zero (e.g., k_0) indicate reference values for the ambient fields. For convenience these are often taken to be the values at a standard height, e.g., 2 m.

The so-called "narrow-angle" PE (valid for propagation angles within about 20° of horizontal), for a 2-D plane in which x is the horizontal coordinate and z vertical, can be written as

$$\frac{\partial A}{\partial x} = \frac{i}{2k_0}\left[\frac{\partial^2}{\partial z^2} + k_{\mathrm{eff}}^2(x,z) - k_0^2\right] A(x,z). \tag{2.6}$$

$A(x, z)$ is the complex carrier amplitude related to the complex sound pressure through $p_c(x, z) = \exp(ik_0 x)A(x, z)$, $k_0 = \omega/c_0$ is a reference value of the wave number, and $k_{\mathrm{eff}} = \omega/c_{\mathrm{eff}}$. The boundary condition along the ground ($z = 0$) is

$$\left.\frac{\partial A(x,z)}{\partial z}\right|_{z=0} = -ik_0\beta A(x,0), \tag{2.7}$$

where $\beta = \rho_0 c_0/Z_s$ is the normalized acoustic admittance of the ground, ρ_0 is the density of the air at $z = 0$, and Z_s is the acoustic ground impedance. Numerical solution of Eqs. (2.6) and (2.7) for a moving medium is essentially the same as for a non-moving medium, because of the validity of the effective sound-speed approximation in the narrow-angle PE.[19]

In some examples below, we will rely on the PE method because it is an effective compromise between physical fidelity and computational cost. Unlike the FFP, the PE method is not restricted to a layered model of the atmosphere and ground, so the sound speed profile and ground impedance can vary along the propagation path. In its broadest form, the PE method also permits accounting for atmospheric turbulence and irregular terrain. The PE can be readily solved using a Crank-Nicholson scheme, the result being the Crank-Nicholson parabolic equation (CNPE) model.[11,21]

2.2.3. *ASL and ground surface modeling considerations*

Consider environmental modeling for input to the sound propagation calculations. Choices made in this portion of the analysis directly impact model error term, ε_m, by bounding the spatial variability that can be accounted for in the near-ground atmosphere through which the sound may undergo refraction and scattering, as well as in the reflection and attenuation of

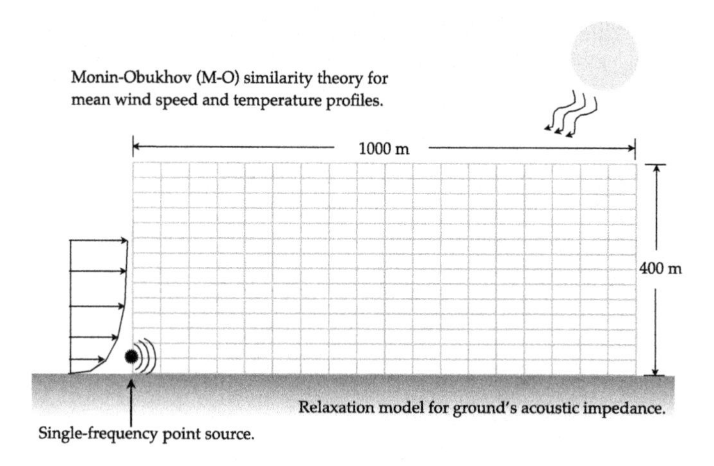

Fig. 2.2. Sketch of two-dimensional sound propagation through the mean atmospheric surface layer.[28] The horizontal and vertical dimensions indicate the extent of the computational grid. This model assumes a flat ground plane.

acoustic energy by the ground; see Fig. 2.2. These choices also influence the other three error budget terms through their conditioning on the chosen model.

In practice, three levels of sophistication are encountered regarding the spatial variability of the ambient atmospheric fields such as c, ρ_{av}, and $\vec{v}_{\mathrm{av}}$: (1) the ambient values may be taken as constants throughout the domain, in which case the sound waves simply spread geometrically through the atmosphere until they encounter the ground or another boundary, (2) only the vertical dependence is considered, or they are taken as constants within predefined horizontal layers of the domain, in which case vertical refraction occurs, or (3) full 2-D or 3-D spatial variability is considered, in which case refraction as well as the turbulent scattering discussed in Sec. 1.1 occurs.

Mean wind and temperature profiles in the ASL are key to sound refraction. The profiles are approximately logarithmic, but deviations occur because of density stratification in the atmosphere. Monin-Obukhov similarity theory (MOST) is widely used in the atmospheric sciences, and increasingly in sound propagation studies,[19,29,30] to model the profiles. MOST parameterizes the profile gradients based on four quantities: height from the ground (z), friction velocity (u_*), surface heat flux (Q_H), and Boussinesq buoyancy parameter (g/T_s), where g is gravitational acceleration and T_s is the temperature at the surface). When the gradients are integrated to determine the actual profiles, a dependence on the surface roughness length (z_0) is introduced. MOST should not be applied above the ASL, and does not hold when turbulent mixing is suppressed, as occurs in strong temperature inversions (usually due to the ground cooling radiatively on a clear night). MOST predicts the *mean* profiles; in reality, the profiles are quite variable due to wind gusts, micro-fronts, and other phenomena. Accounting for unsteady ASL conditions is particularly important in modeling the variability in blast noise propagation; see Secs. 2.5 and 4.1.

Turbulence can be readily handled with the PE and FDTD methods. Perhaps the main challenge of addressing turbulence is not the propagation calculation but the simulation of the turbulent flow. One option is to use a computational fluid dynamics (CFD) simulation. Large-eddy simulation (LES) realistically captures turbulent dynamics in the atmospheric boundary layer, although its resolution (typically no better than a few meters) is often suitable only for sound propagation at low frequencies. Simulations of sound propagation through LES fields can be found in Wilson *et al.*[31]

Turbulence can also be synthesized by Monte Carlo methods, also known as *kinematic turbulence*. The most widely used approach involves realizations of the turbulence produced by randomizing the phases of the modes of the Fourier spectrum and then transforming to the spatial domain.[32,33] To be successful, the kinematic approach requires a realistic turbulence spectrum. The von Kármán spectrum is generally most satisfactory, although it is still idealized in that it does not capture anisotropic structure characteristic of the large eddies. MOST and other turbulence similarity theories can be used to predict parameters in turbulence spectra.[34] The main advantages of the spectral kinematic approaches over CFD are their speed and higher resolution. However, the spectral approaches lack realistic turbulent dynamics, which are important if the goal is to calculate an acoustic time series. Moreover, little research has been done on the impact of kinematic turbulence modeling on skewness, kurtosis, and higher-order turbulence statistics.

Most modeling of outdoor sound propagation in the audible frequency range treats the ground as a rigid-framed porous material; that is, the air within the pores vibrates, but the solid material does not. Viscous and thermal dissipation processes in the pores greatly attenuate the sound. Attenborough *et al.*[12] provide a detailed treatment of this subject. When sound waves in the atmosphere impinge on the ground, the amplitude and phase of the reflection depends greatly on the frequency range and material properties. The volume porosity Ω and static flow resistivity σ are key parameters in most models for the acoustical properties of rigid porous materials. Porosity is defined as the fraction of the porous material occupied by the pore space; the static flow resistivity equals η/k_0, where η is the dynamic viscosity of air and k_0 is the permeability of the porous matrix.

2.3. *Discretization error*

Discretization error seems to receive no special consideration in the computational atmospheric acoustics literature beyond the standard ways of managing convergence and accuracy in all computational solutions of partial differential equations. Section 2.2.2 offers some guidelines for the required spatial and temporal resolutions in terms of the shortest wavelength or highest frequency of interest.

One potential area for future study is the suitability of these resolution guidelines when the atmosphere or ground have strong variations in their acoustic properties with length scales shorter than the smallest wavelength of interest. For example, in simulations of low frequency sound propagating through simulated forest canopies, the required spatial resolution could be set by the spatial variations in the boundary conditions (such as tree trunks and branches), rather than the smallest sound wavelength. Such situations may also occur for sound propagation through turbulence when the primary sound ray is broken up into *correlated micro-paths*, which is known in the wave scattering literature as the *partial saturation* regime.[35]

2.4. *Correctness of probability models*

Probability models of wind speed and direction are essential to doing NGSP UQ, but the difficulty of identifying their proper forms and statistics for a given location or region is a severe impediment to the credibility of high-fidelity spatio-temporal models of NGSP. In wind engineering, probability distributions of near-ground wind speed are commonly assumed to account for the maximum expected wind loads on buildings, bridges, and other structures in the atmospheric surface layer.[36] For example, maximum wind speeds routinely are represented with extreme value statistics derived from on-site measurements or from long-term studies in areas with similar topography, vegetation, soil, and climate. The goal is to have enough data from a long enough period to get the extreme values right.

In NGSP, the concern can be different owing to the potentially greater need for accurate, detailed representations of time-dependent meteorological conditions; see Secs. 2.5 and 4.1. The statistics and perhaps even the most suitable probability distribution of wind speed and direction can vary with location and time, and these factors can affect the accuracy of predictions of frequency-dependence and time-dependence in NGSP. How much this matters depends on the analyst's goal. The details of spatiotemporal variations in meteorological conditions may not matter much if the purpose is to predict the likely range of sound level at a given location or the sensitivity of the sound level to a particular wind speed statistic, but they could be crucial if the purpose is to predict a local community's probability of annoyance under specific blast conditions.

2.5. *Limited data for statistics in probability models*

NGSP studies commonly suffer from inadequate sampling of spatiotemporal variability of wind speeds and temperatures, even in well-controlled field studies. This hinders the use of field studies to validate computational predictions of NGSP under uncertainty. Wilson *et al.*[7] described the resolution of the meteorological state in terms of which type of uncertainty model matches the usual perception of the uncertainty. For example, atmospheric turbulence normally is added to NGSP models as a source of aleatory uncertainty, since the turbulence consists of random and unpredictable motions that are unresolved by atmospheric observations or weather models. Other sources of aleatory uncertainty could be atmospheric motions such as internal gravity waves, small-scale terrain variations, ground impedance, and vegetation, and positions and material properties of buildings. Wilson *et al.* observed that "...the distinction between epistemic and aleatory uncertainty is contingent upon the conceptual model and its parametrization. For example, relatively large-scale atmospheric turbulence could be regarded as an epistemic uncertainty if meteorological towers were employed at many different, closely spaced locations. Then, only the turbulent variations smaller than the space and time scales sampled by the towers would be considered aleatory."

These considerations indicate that field measurements of NGSP to support predictive modeling must be designed with the intended physics model, uncertainty model, and application in mind. It is in general not possible to measure a particular NGSP environment with sufficient spatial and temporal resolution to dependably predict sound propagation as a deterministic process, but mean sound levels may be predictable in certain conditions within acceptable bounds. Therefore, a key consideration in deciding what type of data is needed to support NGSP is whether the goal is to predict a particular sound event or the mean intensity.[34] Sensitivity analysis of the predictive model can support the design of the field study by characterizing the importance of model factors, especially those factors that might be hard to measure well enough in practice.

As discussed in Sec. 2.2.3, MOST is commonly used to model the mean vertical profiles and other turbulence statistics in the ASL. Concerning MOST, there may be substantial epistemic uncertainties in the friction velocity (u_*), wind direction (α), surface heat flux (Q_H), and surface roughness length (z_0). The ground model may furthermore introduce substantial epistemic uncertainties in the static flow resistivity (σ) and porosity (Ω). There are no widely accepted pdfs for these quantities. Ostashev and Wilson[19] used log-normal pdfs for u_*, z_0, and σ, since those quantities are positive definite, and a beta pdf for Ω, since it is bounded between 0 and 1. Normal pdfs can be used for α and Q_H.

3. Uncertainty Quantification and Sensitivity Analysis in Computational NGSP

In this section we summarize recent applications of UQ and SA methods to computational models of NGSP and the related field of underwater sound. Uncertainty quantification methods provide the means to predict the uncertainty in a model's output given uncertainty in its input. Sensitivity analysis methods help to discern which inputs are responsible for the predicted uncertainty in the output. Both UQ and SA methods relate most directly to the $\varepsilon_{p|d,m}$ term in the error budget.

Probability models have appeared in various NGSP models for many years, primarily to address scattering by atmospheric turbulence. For example, in computational models of NGSP, the main method for dealing with the scattering is to generate random realizations of turbulence fields.[32,33] Our focus is broader in seeking practical ways to quantify the effects of all relevant sources of uncertainty in a given application. Our goal is to help ensure recent, generally applicable UQ and SA developments in other fields of computational mechanics are considered for the potential benefits in NGSP predictions.

The UQ discussion is focused on sampling-based methods and stochastic expansions. The SA discussion is focused on two approaches: (1) our recent work in global, full-field sensitivity analysis with probabilistic surrogate models, and (2) variance-based global sensitivity analysis.

This section is intended as a descriptive overview to relate the broad fields of UQ and SA to recent work in computational NGSP predictions. The literature review in this section naturally is biased by our background and approach to the field. In Sec. 4 we offer a broader review organized around specific application contexts.

3.1. *UQ methods in computational NGSP*

In this section we review recent NGSP UQ and SA research with connections to the error budget described in Sec. 2.1.

3.1.1. *Sampling methods in computational NGSP*

The authors and others have used various sampling methods in computational NGSP projects in the past several years. Most recently, Wilson *et al.*[7] examined the effects of various sampling strategies on PE predictions of NGSP through a refracting atmosphere with turbulent scattering. They made the following observations and conclusions:

- When epistemic uncertainties dominate aleatory effects, such as turbulent scattering, Latin hypercube sampling (LHS)[37] can produce more accurate results than ordinary Monte Carlo sampling (MCS).
- Importance sampling, based on understanding of the source spectrum and propagation characteristics can also help by emphasizing samples in the frequency ranges that most strongly contribute to the sound level. This was found to be valuable in avoiding calculations at high frequencies, where solution methods like the PE become very computationally intensive. Some adaptive importance sampling techniques were also considered and found to promote lower calculation times.
- Regardless of the sampling approach, sound levels can generally be predicted more accurately downwind than upwind.
- Calculation times in sampling-based predictions of NGSP variability can be minimized by simultaneously sampling over multiple uncertain variables, as well as variables that are not ordinarily regarded as stochastic, such as frequency and time. Predictions of broadband mean sound levels accurate to 2 to 4 dB out to ranges of 1 km can be obtained using just 16 randomized calculations. The errors were found to decrease approximately as the inverse of the number of realizations. In this way a UQ study of NGSP

calculations can simultaneously incorporate multiple frequencies, turbulent scattering, and uncertainty in atmospheric and ground parameters without substantially increasing the computational burden beyond that of a single broadband calculation involving multiple frequencies, or for a single-frequency calculation that accounts for turbulent scattering through multiple realizations.

3.1.2. *Stochastic expansions in computational NGSP*

The most prominent stochastic expansion method used in computational mechanics studies is polynomial chaos expansion (PCE), but the authors have found no published attempts to use PCE in NGSP. However, Finette and others[2,3,38–41] explored using PCE to simulate stochastic characteristics in the closely related field of ocean acoustics. For example, Finette[2,3] described representing the sound speed and acoustic fields as stochastic processes using PCE. These representations were coupled with a parabolic equation model of sound propagation (see Sec. 2.2.2) to assess statistical moments of the sound field in ocean waveguides.

Truncated PCE were found to accurately reproduce MCS estimates of the mean and variance in the sound pressure field predicted independently with the same physics model. Finette also commented more broadly on the questions about whether adding greater physics fidelity and improving the grid's resolution would should generally improve the quality of ocean acoustics predictions in UQ. Finette[3] observed: "It is possible that embedding a polynomial chaos representation of uncertainty within a relatively simple environmental description of an ocean waveguide may have a better simulation-based predictive capability than a model containing a very detailed environmental representation but without the inclusion of uncertainty in that representation." Although Finette did not state it in these terms, this potential trade-off between physical fidelity and statistical fidelity relates to the error budget discussed in Sec. 2.1. This matter clearly relates to the authors' concern, noted at the end of Sec. 2.1, about wasting computational resources trying to deterministically predict features of the sound field that are dominated by randomness and are therefore unpredictable in practice.

3.2. *SA methods in computational NGSP*

Pettit and Wilson[4] developed full-field representations of the SPL sensitivity to several parameters that define the propagation environment, including the acoustic impedance of the ground and the wind and temperature profiles in the ASL. They referred to this SA approach in later work (e.g., Pettit and Wilson[28]) as *global full-field sensitivity analysis* (GFFSA) because it estimates sensitivities throughout both the physical space of the model (i.e., FFSA) and the parameter space (i.e., GSA).

The GFFSA approach shown in Fig. 2.3 requires (i) sampling throughout the parameter space, generally with LHS or LCVT (Latinized centroidal Voronoi tessellation)[42] to promote uniform sampling, (ii) proper orthogonal decomposition (POD) or independent component analysis (ICA) of the SPL field ensemble produced by the MCS, (iii) regression models of the POD or ICA expansion coefficients, with cluster-weighted modeling (CWM) as the preferred approach, (iv) sensitivity analysis of the regression model of each expansion coefficient, which is very efficient when done for CWM, and (v) projection of the coefficient sensitivities back into the POD or ICA basis to predict SPL sensitivities throughout the spatial domain.

CWM is a form of mixture modeling. Each CWM is a joint pdf of the associated POD or ICA coefficient and the physical parameters. Sampling these distributions throughout the parameter space leads to estimated distributions of the sensitivities, as opposed to only point estimates or mean values. The authors previously used maximum likelihood combined with cross-validation to get point estimates of the CWM parameters and to choose the number of mixture components or clusters. Recently they developed an approximate Bayesian method for inferring the CWM parameters. This helps to regularize the inference process such that cross-validation may be avoided if the number of mixture components is not too high.[43]

The PCE models mentioned in Sec. 3.1.2 allow for sensitivity analysis. The expansion coefficients of the linear terms describe the linear dependence of the SPL on each parameter throughout the spatial domain, so they quantify the full-field first-order sensitivities.

As described above, GSA generalizes the output's local sensitivities to quantify their variations throughout the input space. This is different from the more common variance-based GSA.[44] The authors' GFFSA approach does so through sampling the local analytical derivatives of the expected value of the predictive distributions, each of which is the CWM output distribution of a POD or ICA expansion coefficient conditioned on the inputs. PCE models likely could be adapted to this approach, but the authors are unaware of this being done in acoustics or other areas of computational mechanics. A predictive distribution could be obtained from a PCE model of the sound field, which could then be differentiated analytically and sampled

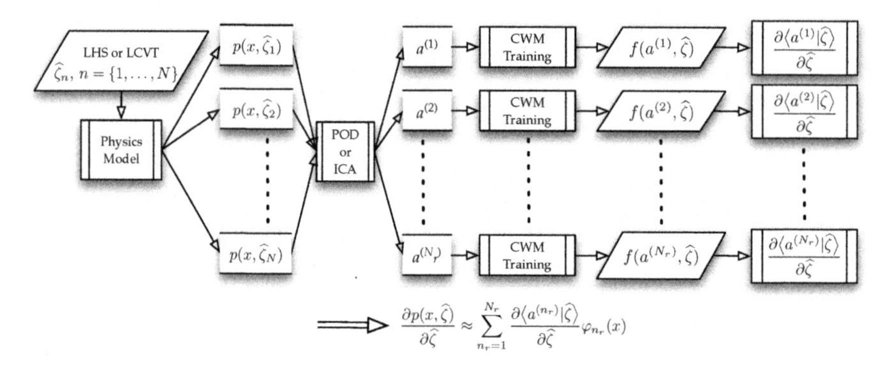

Fig. 2.3. Global, full-field sensitivity analysis. $\widehat{\zeta_n}$ is a vector of uncertain parameters, $p(x, \widehat{\zeta})$ is a snapshot of the random field, $\varphi_{n_r}(x)$ is a POD or ICA mode derived from the snapshots, $a^{(n_r)}$ is a generalized Fourier coefficient of a snapshot relative to the POD or ICA modes, and $f(a^{(n_r)}, \zeta)$ is a joint pdf estimated with a cluster-weighted model. See Pettit and Wilson[28] for further details.

to produce local sensitivities throughout the spatial domain. This would be similar to the authors' GFFSA approach, with the main difference being in how the probabilistic model of the sound field is obtained.

Variance-based GSA forgoes derivatives in favour of estimating the variance of separate terms in a *high-dimensional model representation* (HDMR) based on samples of the model's output throughout the parameter space. Variance-based GSA does not provide the statistical richness of a joint distribution of the inputs and output, but it has the benefit of directly characterizing the degree to which the output's variations in the parameter space depend on each parameter, alone and in various combinations with the other parameters. This could provide unique vantage points on the widely-known trends dating back to the classic paper by Ingard.[9] The authors are unaware of work using variance-based GSA in outdoor acoustics, but this could be done efficiently using either the CWM or PCE described above as surrogate models in place of the full computational model.

4. Application Contexts

In this section, we review some recent research in NGSP with direct connections to UQ and SA. The presentation is arranged to link the applications to specific terms in the error budget described in Sec. 2.1. The featured studies and application areas are not meant to survey the whole field of NGSP modeling; instead, we have two more modest goals: (1) offer some samples of the state of the art in the practical management of uncertainty in NGSP modeling, and (2) demonstrate that the error budget provides a unifying perspective for interpreting the results and benefits of a study, even when the error budget was not considered from the start of the study.

4.1. *Data collection to support uncertainty models*

The second term in the error budget highlights the quality of the data used to populate the distributions upon which UQ is based. In this section we describe recent studies that enrich our understanding of the practical implications of field studies for NGSP simulations. Gauvreau[45] noted that outdoor sound fields always are random across ranges of space and time scales, which "... raises the difficulty of collecting such input data with high-resolution (both in space and time) for use in analytical and numerical models." He described a six-year field study dedicated to providing sufficiently rich data sets in micrometeorology and acoustics to support model validation studies as well as statistical analyses of the spatio-temporal variations in the sound field. He concluded that imprecise measurements and physical variability in near-ground sound studies "... lead to a non-negligible 'random risk' for the experimental results when averaged over 15-min periods. The mean profiles displayed in [their] Fig. 9 should thus be considered with caution."

Junker *et al.*[46] used ray tracing to model NGSP from common environmental noise sources, including roads, railways, and industrial facilities in downward refracting conditions. They also compared their results with earlier PE model simulations. They sought to characterize micrometeorological and ground conditions in terms of effects on sound propagation up to 1 km from the source. The unifying conclusion from this study was that standard statistical treatments of the environmental factors in NGSP are inadequate to support precise predictions. They were able to define required precisions for near-ground meteorological data to distinguish 1 dB(A) variations in predicted relative SPL at the receiver's location. They showed that the required precisions are lower than the accuracy of standard sensors. Moreover, they found that the commonly used definition of an average sound speed profile is questionable because it ignores the high variability in low sound-speed gradient conditions, with the physical effect that small fluctuations in low winds induce substantial sound dispersion.

Juvé[47] developed benchmark lab experiments for validating computational simulations of sound scattering by atmospheric turbulence. Juvé concluded, "The main effects of atmospheric turbulence on sound propagation can be reproduced at laboratory scale, even if complete similarity is not possible." This conclusion was supported by experiments with linear propagation of sound through local and extended turbulent fields, and nonlinear propagation simulating sonic booms.

Valente *et al.*[5] measured blast noise propagation in temperate and desert climates. Their conclusions reinforce the overriding importance of time-dependent meteorological conditions when simulating NGSP, especially for transient events. They observed that instantaneous conditions can be more important than terrain and climate. Moreover, average weather profiles do not suffice for predicting long-range NGSP from short-duration, high-intensity sound, e.g., a blast.

4.2. *Lloyds mirror*

For sound sources above a reflective ground, phase differences between the direct and ground-reflected ray paths produce alternating regions of constructive and destructive interference. This phenomenon, which is known as the acoustical Lloyd's mirror effect,[48] is evident in spectrograms from high-altitude sources, such as aircraft, observed by near-ground listeners.[49] Consider a simple model based on a homogeneous atmosphere to illustrate the effect. The complex sound pressure (as produced by a unit-amplitude source) is modeled with two terms for sound originating from the actual source position and from an image source (the ground-reflected path):

$$p_c(x, z, h, f) = \frac{e^{ikr}}{r} + R_s \frac{e^{ikr'}}{r'}. \tag{2.8}$$

Here, x is the horizontal distance to the receiver (the model has radial symmetry), z is the receiver height, h is the source height, and $r = \sqrt{x^2 + (z-h)^2}$ and $r' = \sqrt{x^2 + (z+h)^2}$ are the distances from the actual and image sources to the observer. Formulas for the spherical wave reflection factor, R_s, are given by Attenborough *et al.*[50] Figure 2.4 shows calculations made with $h = 300\,\text{m}$, $z = 1.5\,\text{m}$, and $f = 1000\,\text{Hz}$. The various curves show the transmission loss (TL) in decibels (dB), defined as $\text{TL} = 20\log_{10}|p_c|$. Predictions for free space ($R_s = 0$), a rigid ground ($R_s = 1$), and an impedance ground, which is characteristic of soil (in which R_s is complex and varies with distance). The Lloyd's mirror effect is evident in the alternating extrema when the rigid or impedance ground is present. The locations of these extrema are very sensitive to the source and receiver heights, the frequency, and the sound speed.

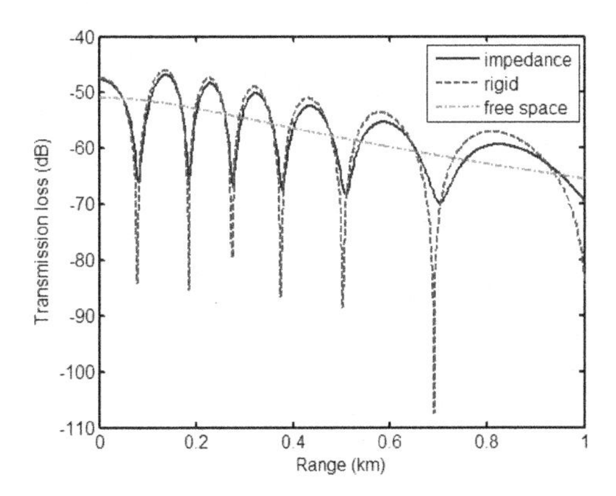

Fig. 2.4. Illustration of the Lloyd's mirror effect.

Consider estimation of the received sound power when there is only a single source of uncertainty, namely in the source height. The total (incoherent) received sound power can be calculated as

$$\Pi = \int_{f_{\min}}^{f_{\max}} \int_{-\infty}^{\infty} S(f) \left|p_c(x, z, h, f)\right|^2 g(h)\, dh\, df, \tag{2.9}$$

where $S(f)$ is the power spectrum of the source (defined between the frequency limits $f_{\min}$ and $f_{\max}$) and $g(h)$ is the pdf for the source height. Generalization of Eq. 2.9 to additional sources of uncertainty is straightforward, by integrating over the additional random variables and replacing $g(h)$ with the joint pdf. The overall SPL in dB at the receiver is

$$\mathrm{SPL}(x, z) = 10 \log_{10} \left[\Pi(x, z)/p_{\mathrm{ref}}^2\right], \tag{2.10}$$

where $p_{\mathrm{ref}} = 20\,\mu\mathrm{Pa}$ is the reference sound pressure.

Wilson *et al.*[7] examined predictability of the SPL in the presence of the Lloyd's mirror effect. A source power spectrum characteristic of an aircraft, which is broadband with a primary peak around 250 Hz, was used, with $h = 300\,\mathrm{m}$ and $z = 1.5\,\mathrm{m}$. Two stochastic approaches to integrating Eq. (2.9) are considered: ordinary MCS and LHS with importance sampling based upon the source spectrum. By varying the number of samples used to evaluate the integrand, and repeating each calculation over 512 trials to obtain an approximate pdf of the prediction error, we determined 90% confidence intervals (CI) for the SPL. Figure 2.5 shows results for a horizontal range of $x = 1000\,\mathrm{m}$. The mean SPL at this range is 80.1 dB. Basic MCS yields a CI spanning about 9 dB for predictions based on 16 integrand

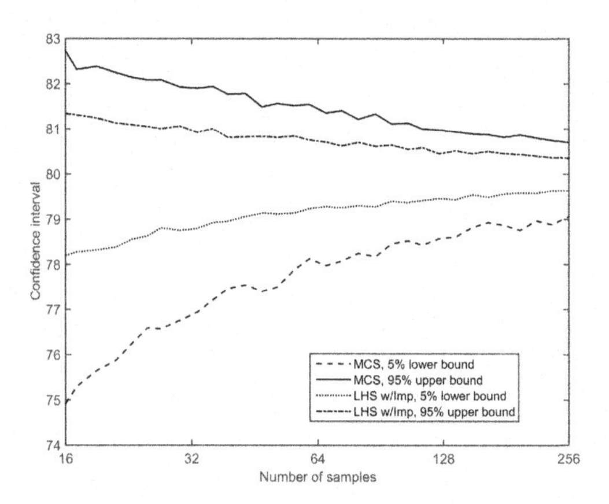

Fig. 2.5. Confidence bounds for estimating the mean SPL at a range of 1 km for MCS and LHS with importance sampling of the source spectrum. Shown are the 90% bounds; i.e., there is a 90% chance an estimate of the mean SPL, made with the given number of samples by the specified sampling method, will lie between the bounds).

samples. The CI decreases roughly as $1/\sqrt{N}$, where N is the number of samples, as would be expected for independent normally distributed samples. LHS with importance sampling shrinks the CI by a factor of nearly 1/3, thus illustrating the potential computational gains in SPL prediction obtainable through careful selection of stochastic sampling methods.

4.3. *Blast noise and battlefield acoustics*

Recent studies have emphasized the centrality of uncertainty in the design of battlefield acoustic sensor networks.[51] The need for better information about the operational setting of the proposed network has grown with the push to integrate higher-fidelity propagation and sensor performance models with optimization methods for choosing sensor locations.

Wilson and Pettit[52] discussed battlefield acoustics (BA) modeling and sensing based on the state-of-the-art predictive skill. They showed BA signal and sensor modeling in five sequential steps. In each step, they highlighted epistemic and aleatory uncertainty and modeling error sources. They emphasized the need to more fully appreciate the essential nature of uncertainty in these predictions, and concluded that increasingly complex modern models for battlefield signal generation and propagation are strongly affected by uncertainty about the environment and the tactical scenario. MCS was used to show this by quantifying how much the probability of detecting the acoustic signatures of air and ground vehicles depended on various levels of uncertainty in the weather and terrain conditions.

This perspective is further supported by the work of Valente *et al.*,[5] already cited in Sec. 4.1. We see from these studies that the conditional dependencies built into the error budget's definition is essential to understanding the combined effects of errors in model physics, measurements, and data. In the case of transient BA simulations, improving the physical fidelity may not be justifiable without more accurately and precisely characterizing the spatio-temporal variations in the environment of interest.

4.4. *Global, full-field sensitivity analysis*

In Sec. 3.2 we described our approach to GFFSA of NGSP simulations based on sampling, POD, and CWM. Figure 2.6 is representative of the results available through this framework.[53] To predict the sensitivities, relative SPL was computed with a wide-angle PE model (see Sec. 2.2) for distances up to 1 km away from a single-frequency, point source near the ground in a steady-state, refracting atmosphere based on MOST (see Sec. 2.2.3). The contour plots show relative levels of predicted linear sensitivity to three of the model's factors within an elevation wedge up to 180 m above the ground at 1 km distance. Results above this wedge were cut off because the wide-angle PE approximation breaks down at higher elevation angles. The plots show that the relative SPL sensitivity is expected to increase with distance from the source and with the source's frequency.

The results in Fig. 2.6 must be interpreted within the underlying modeling restrictions. For example, they come from a simulated ensemble on samples throughout the space of model factors. In each realization, the ASL and ground impedance factors are constant throughout the physical domain, so they vary only from one realization to the next. No attempt is made to include turbulent scattering. In this sense, the results of our GFFSA work to date are relevant and make physical sense within the limitations of the assumptions (as described in Pettit and Wilson[4]), but they are primarily a proof of concept. The state of the art could be advanced by applying GFFSA to higher fidelity models featuring time-dependent turbulence in the ASL, such as those examined by the authors and others.[31]

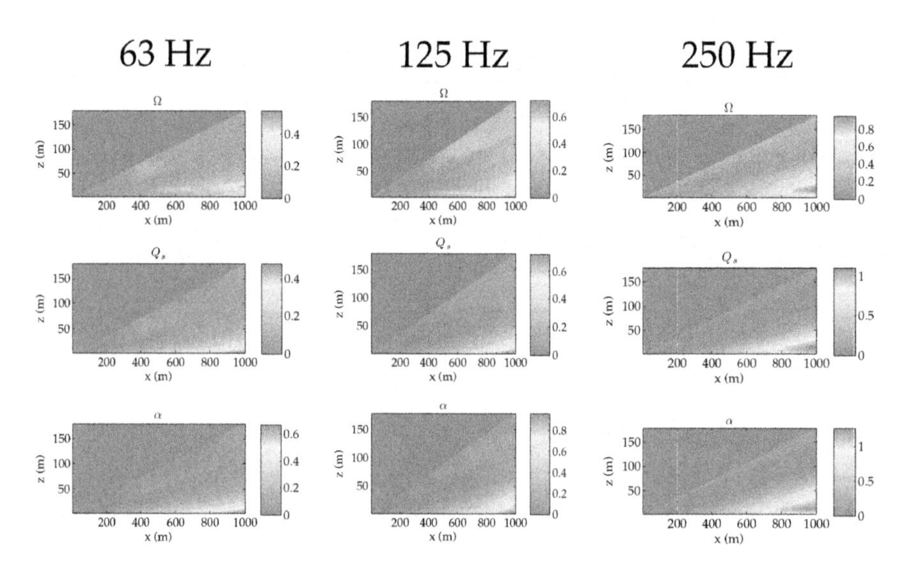

Fig. 2.6. Ensemble averages of the normalized first-order sensitivities of relative sound pressure level to soil porosity Ω, surface heat flux Q_S, and wind direction α. SPL is due to a monopole source at $x = 0$ with the frequencies shown. Absolute values are shown to compensate for the effect of wind direction.[53]

5. Conclusions

Near-ground sound propagation (NGSP) is sensitive to many factors that vary in time and in space. High-fidelity computational models are available, but each model brings with it a unique set of errors as well as uncertainty about the data that support its parametrization. In the most sophisticated models, accurately characterizing the spatial and temporal features of the environment at the small scales necessary to fully utilize their capabilities may be impractical. Therefore, a primary concern in the field of near-ground sound propagation is optimizing the combination of physics models, field data, and computing resources.

We promoted an *error budget* to itemize sources of predictive error and to discern how they combine to restrict predictive skill. We reviewed the recent literature in Monte Carlo simulation, uncertainty quantification, and sensitivity analysis of NGSP to characterize the NGSP error budget. We strove to highlight specific error budget terms and applications in which uncertainty and errors noticeably restrict the ability to predict NGSP with high confidence. For example, we found that the relative efficiency of a sampling method depends on whether uncertainties in the mean properties of the atmosphere and ground dominate, or random turbulent scattering dominates. This relates to managing the third term in the error budget we described. We encourage researchers to incorporate the error budget in studies directed at advancing the state of the art in predicting NGSP.

Acknowledgement

This research was supported by the U.S. Army Engineer Research and Development Center (ERDC) Geospatial Research and Engineering business area. Permission to publish was granted by Director, Cold Regions Research and Engineering Laboratory (CRREL).

References

1. S. E. Dosso and P. L. Nielsen, Quantifying uncertainty in geoacoustic inversion. I. a fast Gibbs sampler approach, *J. Acoust. Soc. Am.* **111**(1), 129–142 (2002).
2. S. Finette, Embedding uncertainty into ocean acoustic propagation models, *J. Acoust. Soc. Am.* **117**(3), 997–1000 (2005).
3. S. Finette, A stochastic representation of environmental uncertainty and its coupling to acoustic wave propagation in ocean waveguides, *J. Acoust. Soc. Am.* **120**(5), 2567–2579 (2006).
4. C. L. Pettit and D. K. Wilson, Proper orthogonal decomposition and cluster weighted modeling for sensitivity analysis of sound propagation in the atmospheric surface layer, *J. Acoust. Soc. Am.* **122**(3), 1374–1390 (2007).
5. D. Valente, L. M. Ronsse, L. Pater, M. J. White, R. Serwy, E. T. Nykaza, M. E. Swearingen and D. G. Albert, Blast noise characteristics as a function of distance for temperate and desert climates, *J. Acoust. Soc. Am.* **132**(1), 216–227 (2012).
6. R. Ghanem, Error budgets: A path from uncertainty quantification to model validation. In *Advanced Simulation and Computing Workshop: Error Estimation, Uncertainty Quantification, and Reliability in Numerical Simulations*, Stanford, CA (August, 2005).
7. D. K. Wilson, C. L. Pettit, V. E. Ostashev and S. N. Vecherin, Description and quantification of uncertainty in outdoor sound propagation calculations, *J. Acoust. Soc. Am.* **136**(3), 1013–1028 (2014).
8. R. B. Stull, *An Introduction to Boundary Layer Meteorology.* Springer (1988).
9. U. Ingärd, A review of the influence of meteorological conditions on sound propagation, *J. Acoust. Soc. Am.* **25**(3), 405–411 (1953).
10. T. F. W. Embleton, Tutorial on sound propagation outdoors, *J. Acoust. Soc. Am.* **100**(1), 31–48 (1996).
11. E. M. Salomons, *Computational Atmospheric Acoustics.* Kluwer Academic Publishers, Dordrecht, The Netherlands (2001).

12. K. Attenborough, K. M. Li, and K. Horoshenkov, *Predicting Outdoor Sound*. Taylor & Francis (2007).
13. D. K. Wilson, Simple, relaxational models for the acoustical properties of porous media, *Appl. Acoust.* **50**, 171–188 (1997).
14. M. T. Reagan, H. N. Najm, R. G. Ghanem, and O. M. Knio, Uncertainty quantification in reacting-flow simulations through non-intrusive spectral projection, *Combust. Flame.* **132**, 545–555 (2003).
15. A. Desai and S. Sarkar, Analysis of a nonlinear aeroelastic system with parametric uncertainties using polynomial chaos expansion, *Math. Problems Eng.* pp. 1–21 (2010), doi:10.1155/2010/379472.
16. O. Leroy, B. Gauvreau, F. Junker, E. de Rocquigny and M. Bérengier, Uncertainty assessment for outdoor sound propagation. In *20th International Congress on Acoustics*, pp. 192–198, Sydney, Australia (August, 2010).
17. C. L. Pettit and P. S. Beran. Convergence studies of wiener expansions for computational nonlinear mechanics. In *8th AIAA Non-Deterministic Approaches Conference*, number AIAA-2006-1993, Newport, RI (May, 2006).
18. A. D. Pierce, *Acoustics: An Introduction to its Physical Principles and Applications*. McGraw-Hill (1989).
19. V. E. Ostashev and D. K. Wilson, *Acoustics in Moving Inhomogeneous Media*, second edn. CRC Press, Boca Raton, FL (2015).
20. K. E. Gilbert and M. J. White, Application of the parabolic equation to sound propagation in a refracting atmosphere, *J. Acoust. Soc. Am.* **85**, 630–637 (1989).
21. M. West, K. Gilbert, and R. A. Sack, A tutorial on the parabolic equation (PE) model used for long range sound propagation in the atmosphere, *Appl. Acoust.* **37**(1), 31–49 (1992).
22. S. Lee, N. Bong, W. Richards, and R. Raspet, Impedance formulation of the fast field program for acoustic wave propagation in the atmosphere, *J. Acoust. Soc. Am.* **79**, 628–634 (1986).
23. D. K. Wilson, Sound field computations in a stratified, moving medium, *J. Acoust. Soc. Am.* **94**, 400–407 (1993).
24. R. Blumrich and D. Heimann, A linearized Eulerian sound propagation model for studies of complex meteorological effects, *J. Acoust. Soc. Am.* **112**(2), 446–455 (2002).
25. V. E. Ostashev, D. K. Wilson, L. Liu, D. F. Aldridge, N. P. Symons, and D. Marlin, Equations for finite-difference, time-domain simulation of sound propagation in moving inhomogeneous media and numerical implementation, *J. Acoust. Soc. Am.* **117**, 503–517 (2005).
26. M. Hornikx, R. Waxler, and J. Forssén, The extended Fourier pseudospectral time-domain method for atmospheric sound propagation, *J. Acoust. Soc. Am.* **128**(4), 1632–1646 (2010).
27. E. Premat and Y. Gabillet, A new boundary-element method for predicting outdoor sound propagation and application to the case of a sound barrier in the presence of downward refraction, *J. Acoust. Soc. Am.* **108**(6), 2775–2783 (2000).

28. C. L. Pettit and D. K. Wilson, Full-field sensitivity analysis through dimension reduction and probabilistic surrogate models, *Probab. Eng. Mech.* **25**(4), 380–392 (2010).
29. D. K. Wilson, The sound-speed gradient and refraction in the near-ground atmosphere, *J. Acoust. Soc. Am.* **113**(2), 750–757 (2003).
30. D. Heimann and E. Salomons, Testing meteorological classifications for the prediction of long-term average sound levels, *Appl. Acoust.* **65**(10), 925–950 (2004).
31. D. K. Wilson, E. L. Andreas, J. W. Weatherly, C. L. Pettit, E. G. Patton, and P. P. Sullivan, Characterization of uncertainty in outdoor sound propagation predictions, *J. Acoust. Soc. Am.* **121**(5), EL177–EL183 (2007).
32. P. Chevret, P. Blanc-Benon and D. Juvé, A numerical model for sound propagation through a turbulent atmosphere near the ground, *J. Acoust. Soc. Am.* **100**, 3587–3599 (1996).
33. K. E. Gilbert, R. Raspet and X. Di, Calculation of turbulence effects in an upward-refracting atmosphere, *J. Acoust. Soc. Am.* **87**, 2428–2437 (1990).
34. D. K. Wilson, M. S. Lewis, J. W. Weatherly and E. L. Andreas, Dependence of predictive skill for outdoor narrowband and broadband sound levels on the atmospheric representation, *Noise Control Eng. J.* **56**(6), 465–477 (2008).
35. S. M. Flatté, R. Dashen, W. H. Munk, K. M. Watson, and F. Zachariasen, *Sound Transmission Through a Fluctuating Ocean.* Cambridge University Press, Cambridge (1979).
36. E. Simiu and R. H. Scanlan, *Wind Effects on Structures*, 3rd edn. Wiley (1996).
37. J. C. Helton and F. J. Davis, Latin hypercube sampling and the propagation of uncertainty in analyses of complex systems, *Reliab. Eng. Syst. Saf.* **81**, 23–69 (2003).
38. D. B. Creamer, On using polynomial chaos for modeling uncertainty in acoustic propagation, *J. Acoust. Soc. Am.* **119**(4), 1979–1994 (2006).
39. K. D. LePage, Estimation of acoustic propagation uncertainty through polynomial chaos expansions. In *2006 9th International Conference on Information Fusion*, pp. 1–5 (2006).
40. S. Finette, A stochastic response surface formulation of acoustic propagation through an uncertain ocean waveguide environment, *J. Acoust. Soc. Am.* **126** (5), 2242–2247 (2009).
41. Y. Y. Khine, D. B. Creamer, and S. Finette, Acoustic propagation in an uncertain waveguide environment using stochastic basis expansions, *J. Comput. Acoust.* **18**(04), 397–441 (2010).
42. J. Burkardt, Latin centroidal Voronoi tessellations, Webpage (May, 2007).
43. C. L. Pettit and D. K. Wilson, Variational inference of cluster-weighted models for local and global sensitivity analysis, *Int. J. Reliab. Saf.* (2014 (accepted for publication)).
44. A. Saltelli, M. Ratto, T. Andres, and F. Campolongo, *Global Sensitivity Analysis: The Primer.* Wiley, Chichester (2008).
45. B. Gauvreau, Long-term experimental database for environmental acoustics, *Appl. Acoust.* **74**(7), 958 – 967 (2013).

46. F. Junker, B. Bauvreau, D. Ecotié, C. Crémézi-Charlet and P. Blanc-Benon, Meteorological classification for environmental acoustics — practical implications due to experimental accuracy and uncertainty. In *19th International Congress on Acoustics*, pp. 2882–2887, Madrid, Spain (September, 2007).
47. D. Juvé, Juve-2013-experimental and numerical simulation of atmospheric sound propagation: effects of turbulence and nonlinearities.pdf. In *17th CEAS-ASC and 3rd X-Noise EV Workshop*, Sevilla (September, 2013).
48. W. M. Carey, Lloyd's mirror-image interference effects, *Acoust. Today* **5**(2), 14–20 (2009).
49. K. W. Lo, S. W. Perry, and B. G. Ferguson, Aircraft flight parameter estimation using acoustical Lloyd's mirror effect, *IEEE Trans. Aerospace Electron. Syst.* **38**(1), 137–151 (2002).
50. K. Attenborough, S. I. Hayek, and J. M. Lawther, Propagation of sound over a porous half-space. *J. Acoust. Soc. Am.* **68**(5), 1493–1501 (1980).
51. S. N. Vecherin, D. K. Wilson and C. L. Pettit, Optimal placement of multimodal sensors subject to supply, communication, and robustness constraints. In *Joint Meeting of the Military Sensing Symposia Specialty Group on BAMS*, Laurel, MD (September, 2010).
52. D. K. Wilson and C. L. Pettit, Uncertainty in battlefield acoustics predictions. In *Joint Meeting of the Military Sensing Symposia Specialty Group on BAMS*, Laurel, MD (September, 2010).
53. C. L. Pettit and D. K. Wilson, Variational inference of global sensitivities in near-ground sound propagation. In *14th Long Range Sound Propagation Symposium*, Annapolis, MD, USA (March, 2011).

Chapter 3

Efficient Uncertainty Analysis of Radiative Heating for Planetary Entry

Thomas K. West IV* and Serhat Hosder

Department of Mechanical and Aerospace Engineering
Missouri University of Science and Technology
1870 Miner Circle, Rolla, MO 65409, USA
**tkwgg3@mst.edu*

Computational fluid dynamics simulations of hypersonic, planetary entry flows and radiative heating predictions possess a significant amount of uncertainty due to the complexity of the flow physics and the difficulty in obtaining accurate experimental results of molecular level phenomena. In addition, the complexity of the flow physics requires high-fidelity, numerical models, which are computationally expensive. Therefore, quantifying the uncertainty in such models with classical sampling approaches becomes infeasible due to the large number of model evaluations required to obtain the statistics with desired accuracy. In this chapter, a computationally efficient means of uncertainty quantification was introduced and demonstrated for the prediction of radiative heating during Mars entry. The approach was to construct a surrogate model using a sparse approximation of the point-collocation non-intrusive polynomial chaos method. While polynomial chaos methods suffer from the curse of dimensionality, the sparse approximation method alleviates the cost for large-scale problems, such as the high-fidelity numerical modelling of planetary entry flows with radiative heating. The results show that an accurate stochastic surrogate model could be constructed with only 500 samples of the computational model. This is about 10% of the cost to construct the same surrogate model without the sparse approximation and corresponds to a significantly less number of samples than required for a pure sampling-based uncertainty quantification approach.

1. Introduction

In the present day, numerical modelling of complex physics problems has become one of the important tools in the analysis and design of aerospace systems. One example is the analysis of hypersonic, atmospheric entry flows and the design of reliable thermal protection systems (TPS). The analysis of these particular flows is difficult due to the complexity of the physics governing, not only the fluid dynamics, but also the thermodynamics, heat transfer, and flow field chemical kinetics. With recent advancements in computational hardware and numerical algorithms, computational fluid dynamics (CFD) has become an attractive approach for modelling these complex physical phenomena.

Because of the complex flow physics, a significant amount of uncertainty may exist in the modelling of planetary entry flows. This makes accurate heat flux predictions a challenge and emphasizes the need for quantification of this uncertainty. Due to the computational expense of the numerical simulations of planetary entry flows, performing uncertainty quantification (UQ) with traditional sampling approaches may not be feasible. Surrogate modelling may be seen as an alternative means of efficient UQ; however, many basic surrogate modelling approaches suffer from a "curse of dimensionality" in that the number of required evaluations of the computationally expensive model grows exponentially with the number of random variables. As a results, a new approach is needed to allow for efficient and accurate UQ of large scale, complex problems.

The primary objective of this chapter is to address these challenges by outlining and demonstrating a computationally efficient surrogate modelling approach for use as an efficient means of UQ. This will be achieved by constructing a polynomial chaos expansion (PCE) through the use of a sparse approximation of the point-collocation nonintrusive polynomial chaos (NIPC) method. The constructed surrogate model will replace a computationally expensive model used for the prediction of radiative heating during entry into the Mars atmosphere. The goal is to minimize the number of computationally expensive deterministic model evaluations needed for an accurate UQ analysis.

The following section outlines the uncertainty quantification methodology based on a sparse approximation of the PCE. Section 3 demonstrates the approach on an analytical function. The next section then demonstrates the sparse approximation technique on a high-fidelity, CFD model used for predicting radiative heat flux on a spacecraft during entry into the Mars atmosphere. Details regarding the CFD model, entry conditions, and discussion of baseline solutions are also given in this section. The final section gives a summary of the chapter.

2. Uncertainty Quantification Approach

This section provides the details of the surrogate modelling and uncertainty quantification approach. The first part outlines the general non-intrusive polynomial chaos formulation with the point-collocation approach. The second part details the solution recovery approach for determining the PCE coefficients under sparse conditions, followed by a discussion of error and convergences measures. Lastly, a discussion of the types of uncertainty in numerical modelling is provided.

2.1. *Point-collocation non-intrusive polynomial chaos*

In recent studies,[1–5] the polynomial chaos method has been used as a means of UQ over traditional methods, such as Monte Carlo, for computational efficiency. Polynomial chaos can be viewed as a surrogate modelling technique based on a spectral representation of uncertainty.[6] In the spectral representation, an uncertain response value or random function, α^*, is decomposed into separable deterministic and stochastic components, as shown in Eq. (3.1).

$$\alpha^*(\mathbf{x}, \boldsymbol{\xi}) = \sum_{i=0}^{\infty} \alpha_i(\mathbf{x})\Psi_i(\boldsymbol{\xi}). \tag{3.1}$$

Here, α_i is the deterministic component and Ψ_i is the random component. α^* is function of deterministic variables, $\mathbf{x}$, and n-dimensional standard random variables, $\boldsymbol{\xi}$. In polynomial chaos, the behaviour of random variables within their prescribed domain is assumed to be known. The random component Ψ_i is then prescribed based on the distribution of the random variables. The Wiener-Askey scheme[7] provides a set of polynomials that provide an optimal basis for various continuous probability density functions. For example, the Legendre polynomials provide an optimal basis for a uniform probability density function.

By definition, the expansion given by Eq. (3.1) is an infinite series. In engineering practice, this series is truncated to a finite number of terms, as given by Eq. (3.2). To form a complete basis or for a total order expansion,

N_t terms are required, which can be computed from Eq. (3.3) for a PCE of order p and a number of random dimensions or variables, n.

$$\alpha^*(\mathbf{x}, \boldsymbol{\xi}) \approx \sum_{i=0}^{P} \alpha_i(\mathbf{x}) \Psi_i(\boldsymbol{\xi}), \tag{3.2}$$

$$N_t = P + 1 = \frac{(n+p)!}{n!p!}. \tag{3.3}$$

The objective with any PCE method is to determine the expansion coefficients, α_i. To do this, polynomial chaos methods can be implemented using an intrusive or a non-intrusive approach. While an intrusive method may appear straightforward in theory, for complex problems this process may be time consuming, expensive, and difficult to implement. In contrast, the non-intrusive polynomial chaos (NIPC) approach can be easily implemented to construct a surrogate model that represents a complex computational simulation, because no modification to the deterministic model is required. The non-intrusive methods require only the response values at selected sample points to approximate the stochastic response surface.

Several methods have been developed for NIPC. The reader is encouraged to survey the literature to learn other NIPC approaches (see work by Eldred[8]) and view the wide array of engineering problems for which they have been applied. Of these existing methods, the point-collocation NIPC method has been used extensively in many aerospace applications.[2,3,5,9] This approach starts with replacing a stochastic response or random function with its PCE by using Eq. (3.2). Then, N_t sample vectors are chosen in random space and the deterministic code is evaluated at these points, which is the left hand side of Eq. (3.2). Following this, a linear system of N_t equations can be formulated and solved for the spectral modes of the random variables. This system is shown as follows:

$$\begin{pmatrix} \alpha^*(\mathbf{x}, \boldsymbol{\xi}_0) \\ \alpha^*(\mathbf{x}, \boldsymbol{\xi}_1) \\ \vdots \\ \alpha^*(\mathbf{x}, \boldsymbol{\xi}_P) \end{pmatrix} = \begin{pmatrix} \Psi_0(\boldsymbol{\xi}_0) & \Psi_1(\boldsymbol{\xi}_0) & \cdots & \Psi_P(\boldsymbol{\xi}_0) \\ \Psi_0(\boldsymbol{\xi}_1) & \Psi_1(\boldsymbol{\xi}_1) & \cdots & \Psi_P(\boldsymbol{\xi}_1) \\ \vdots & \vdots & \ddots & \vdots \\ \Psi_0(\boldsymbol{\xi}_P) & \Psi_1(\boldsymbol{\xi}_P) & \cdots & \Psi_P(\boldsymbol{\xi}_P) \end{pmatrix} \begin{pmatrix} \alpha_0 \\ \alpha_1 \\ \vdots \\ \alpha_P \end{pmatrix}. \tag{3.4}$$

Note that for this linear system, N_t is the minimum number of samples required to solve the determined system (i.e., the coefficient vector). If more samples are available and are linearly independent, the system is considered overdetermined and can be solved using a least squares approach. A previous work has shown that oversampling by a factor of two may improve the accuracy of the PCE.[10]

Polynomial chaos techniques suffer from a "curse of dimensionality". This means that the number of deterministic model evaluations required to create an accurate surrogate model grows exponentially with the number of random dimensions. For many large-scale, complex problems, such as those found in modelling hypersonic, planetary entry flows, obtaining the minimum number of deterministic model samples required to construct a PCE may be infeasible or even impossible. The most desirable approach is then to obtain an accurate surrogate model with as few deterministic samples as possible to limit the computational cost, even if the minimum number of samples required for a total order expansion is not achievable.

2.2. *Solution approach with a sparse approximation*

In general, a system of linear equations, like the one shown in Eq. (3.4), that has fewer linearly independent equations than unknowns, possesses an infinite number of solutions. In many PCEs, only a small fraction of the coefficients may carry significant weight on the surrogate model. This would then allow for an assumption that many of the expansion coefficients are zero, making the vector of expansion coefficients sparse. With this assumption, a linear system can be regularized allowing for a well-posed solution. The objective is to seek out a solution to the linear system with the fewest number of non-zero coefficients. Using convex relaxation, a solution can be obtained from the L_1-minimization problem shown in Eq. (3.5).

$$\min \left\|\alpha\right\|_1 \text{ subject to } \left\|\Psi\alpha - \alpha^*\right\|_2 \leq \delta. \tag{3.5}$$

Here, δ is the error associated with the truncation of the series in Eq. (3.2). There are multiple ways to obtain the best estimate for this value. One approach is to simply assume $\delta = 0$, as this case can be shown to produce a unique solution to Eq. (3.5). In the above formulation, the dimensions of Ψ are N_s x N_t and the vector α^* is of length N_s, where the number of samples, N_s, is less than the number of terms, N_t, for the underdetermined problem. The vector α is of length N_t. Note that there is no dependence or restriction on the order of the PCE approximation. The reader is directed to the literature, such as the work by Doostan and Owhadi,[11] for additional discussion of the mathematical formulation and stability.

The optimization problem in Eq. (3.5) is commonly referred to as Basis Pursuit Denoising (BPDN). These types of problems can be solved using many methods from quadratic programming, and the discussion of these methods are left to other works, such as those by Yang *et al.*[12] and Asif *et al.*[13] The technique employed here is the least absolute shrinkage and selection operator (LASSO) homotopy optimization routine.[13] While many methods exist for solving the above minimization problem, the homotopy method was selected for efficiency, as this method is not significantly affected by the dimensionality of the problem.[12] The downside to this approach is that for problems with a sparseness ratio approaching one, the optimization time may increase; however, the expectation is that for problems with a large number of random variables, only a few will be significant making the homotopy approach ideal for solving these problems.

2.3. *Sample size, accuracy, and convergence*

The optimization and sparse solution recovery approach poses two fundamental issues: (1) how to determine the necessary number of samples, N_s, required to obtain an accurate solution and (2) how to measure the accuracy of the solution. The latter of these assumes, of course, that no other means of obtaining the exact solution is possible, thereby relying on the solution obtained from Eq. (3.5).

To reduce the computational cost, the desired approach is to limit the total number of deterministic model evaluations. One approach to this is to iteratively increase the size of the sample set used to recover the sparse approximation (i.e., solve Eq. (3.5)). While many approaches to iteratively increasing the sample set exist and/or can be formulated, for simplicity, the approach used here is based on the idea of randomly picking subsets from a set of samples equal to the number of terms, N_t, in the PCE. A Latin Hypercube sample structure can be used for adequate coverage of the domain spanned by the uncertain parameters.

To start, the deterministic model is evaluated at the points of a small subset of the initial sample structure. Then, a first set of PCE coefficients can be obtained using the minimization routine in Eq. (3.5). This process is then repeated by iteratively adding more samples to the solution procedure (i.e., addition of new subsets of the full sample structure) until adequate convergence of the expansion coefficients is achieved. Note that each subset of the full sample structure added at each iteration should not contain any repeated sample vectors from the previous iterations because this would not provide any new information in recovering new solutions at each iteration. Also, as more samples are added to the structure used to recover the sparse approximation, the degree of sparseness will likely decrease meaning that more of the coefficients will become nonzero. This is expected as more information may warrant some coefficients being nonzero to improve the sparse approximation (i.e., satisfy the optimization constraint).

After the expansion coefficients are approximated, their convergence should be checked at each iteration. In theory, this could be done by monitoring the convergence of each coefficient. Unfortunately, for large scale problems, there may be thousands of coefficients. Also, because the expansion coefficient vector is known to be sparse, this may not be an accurate approach as the degree of sparseness of the solution vector may decrease with increasing sample size causing radical changes in any convergence error measurement. A logical choice for a convergence metric would be to use output statistics based on the expansion coefficients. One approach is to use what are known as Sobol indices.[14] Sobol indices can be derived via *Sobol Decomposition*, which is a variance-based global sensitivity analysis method. Sobol indices have the distinct advantage of providing sensitivity information regarding each of the uncertain parameters and is an important tool for determining which parameters contribute most significantly to the output variance. For completeness, a detailed description of the Sobol indices is given here.

First, the total variance, D, can be written in terms of the PCE as shown in Eq. (3.6).

$$D = \sum_{j=1}^{P} \alpha_j^2(t, \vec{x}) \langle \Psi_j^2(\vec{\xi}) \rangle. \tag{3.6}$$

Then, the total variance can be decomposed as:

$$D = \sum_{i=1}^{i=n} D_i + \sum_{1 \le i < j \le n}^{i=n-1} D_{i,j} + \sum_{1 \le i < j < k \le n}^{i=n-2} D_{i,j,k} + \cdots + D_{1,2,\ldots,n}, \tag{3.7}$$

where the partial variances ($D_{i_1,\ldots,i_s}$) are given by:

$$D_{i_1,\ldots,i_s} = \sum_{\beta \in \{i_1,\ldots,i_s\}} \alpha_\beta^2 \langle \Psi_\beta^2(\vec{\xi}) \rangle, \qquad 1 \le i_1 < \ldots < i_s \le n. \tag{3.8}$$

Then the Sobol indices ($S_{i_1 \cdots i_s}$) are defined as,

$$S_{i_1 \cdots i_s} = \frac{D_{i_1,\ldots,i_s}}{D}, \tag{3.9}$$

which satisfy the following equation:

$$\sum_{i=1}^{i=n} S_i + \sum_{1 \leq i<j \leq n}^{i=n-1} S_{i,j} + \sum_{1 \leq i<j<k \leq n}^{i=n-2} S_{i,j,k} + \cdots + S_{1,2,\ldots,n} = 1.0. \quad (3.10)$$

The Sobol indices provide a sensitivity measure due to individual contribution from each input uncertain variable (S_i), as well as the mixed contributions ($\{S_{i,j}\}, \{S_{i,j,k}\}, \ldots$). The total (combined) effect (S_{T_i}) of an input parameter i is defined as the summation of the partial Sobol indices that include the particular parameter:

$$S_{T_i} = \sum_{L_i} \frac{D_{i_1,\ldots,i_s}}{D}; \quad L_i = \{(i_1, \ldots, i_s) : \exists k,\ 1 \leq k \leq s,\ i_k = i\}. \quad (3.11)$$

For example, with $n = 3$, the total contribution to the overall variance from the first uncertain variable ($i = 1$) can be written as:

$$S_{T_1} = S_1 + S_{1,2} + S_{1,3} + S_{1,2,3}. \quad (3.12)$$

From these formulations, the Sobol indices can be used to provide a relative ranking of each input uncertainty to the overall variation in the output with the consideration of nonlinear correlation between input variables and output quantities of interest.

The accuracy of the Sobol indices depend highly on the accuracy of the PCE coefficients, making it an ideal measure of their convergence. Also, because the number of total Sobol indices is the same as the number of uncertain parameters, there is less parameters to track, as this number will always be less than the number of PCE coefficients. To monitor the convergence of the total Sobol indices with the addition of more samples at each iteration, an absolute error, $S_{e_{i,j}}$ can be defined for the jth total Sobol index at iteration i using Eq. (3.13).

$$S_{e_{i,j}} = \left\| S_{T,i,j} - S_{T,i-1,j} \right\|. \quad (3.13)$$

Note that measuring the convergence based on this absolute error puts emphasis on the variables that contribute more to the output uncertainty. The error of each total Sobol index, at each iteration, can then be averaged giving a single value for monitoring, which is shown with Eq. (3.14).

$$\mu_{e,i} = \frac{1}{n} \sum_{j=1}^{n} S_{e_{i,j}}. \tag{3.14}$$

Tracking this average error at each iteration would then illustrate the convergence of the PCE coefficients. The objective will be to seek out nearly asymptotic convergence, as zero error would likely not be achievable simply due to the randomness of the samples added at each iteration and any numerical inaccuracies that may occur during the analysis of complex models.

Tracking the convergence of the Sobol indices addresses the question of how many samples are necessary for the convergence of the PCE coefficients, but does not provide a measure of the accuracy of the recovered solution. This can be done by comparing the response values obtained with the PCE to the actual response values at selected sample locations that are different than the ones used to train the stochastic surrogate. The objective is to measure the error between response values from the polynomial approximation and the actual response at these test sample locations. The test points can be chosen throughout the design space using a Latin Hypercube structure. This has the advantage of improving the coverage of the design space when a small sample set is used. Note that each test point used to measure the accuracy of the surrogate model does require an additional evaluation of the deterministic model to obtain the exact functional value. These test points could be rolled into the sample set used to approximate the surrogate, which could improve the accuracy of the model as more samples are included in the solution recovery. The magnitude of improvement will depend on the convergence (i.e., the accuracy of the model at the current iteration) and the number of test points used.

A measure of the response surface error at each iteration, $T_{e,i}$, may be estimated by the mean square error defined by Eq. (3.15).

$$T_{e,i} = \frac{1}{N_{TP}} \sum_{j=1}^{N_{TP}} (F_{surr.}(\mathbf{x_j}) - F_{actual}(\mathbf{x_j}))^2. \tag{3.15}$$

Here, N_{TP} is the number of test points, $F_{surr.}$ is the response value from the PCE surrogate model, and F_{actual} is the actual test point value from the design space. The error between the surrogate and the deterministic model, at each test point, is an indication of the local accuracy of the surrogate model. Maximizing the number of test points will provide better coverage in the design space and would provide the best indication of the accuracy of the surrogate. Again, however, this does come at the cost of additional evaluations of the deterministic model.

2.4. *Types of uncertainty and uncertainty propagation*

Two main types of uncertainty exist in numerical modelling: aleatory uncertainty and epistemic uncertainty.[15] Aleatory uncertainty is the inherent variation of a physical system. Such variation is due the random nature of input data and can be mathematically represented by a probability density function if substantial experimental data is available for estimating the distribution type. An example of this for stochastic CFD simulations could be the fluctuation in freestream quantities. While still considered a random variable, these variables are not controllable and their uncertainty is sometimes referred to as irreducible.

After constructing the stochastic response surface, propagation of aleatory uncertainty can be accomplished with a Monte Carlo analysis. While large sample size are typically needed for accurate results (on the order of 10^5), sampling the surrogate model is extremely inexpensive compared to sampling the actual deterministic model. With the aleatory uncertainty analysis, a cumulative distribution function can be constructed from the outputs and various statistics can be obtained, such as, for examples, confidence intervals.

Epistemic uncertainty in a stochastic problem comes from several potential sources. These include a lack of knowledge or incomplete information of the behaviour of a particular variable and ignorance or negligence with regards to accurate modelling of model parameters. Contrary to aleatory uncertainty, epistemic uncertainty is sometimes referred to as reducible uncertainty as an increase in knowledge regarding the physics of a problem, along with accurate modelling, can reduce the amount of this type of uncertainty. Epistemic uncertainty is typically modelled using intervals because the use of probabilistic distributions can lead to inaccurate predictions in the amount of uncertainty in a system. Upper and lower bounds of these intervals can be drawn from limited experimental data or from expert predictions and judgment.

Propagating input, epistemic uncertainty can also be accomplished using a Monte Carlo sampling approach to obtain an output/ response interval. No probability function can be constructed from the outputs in this case, however. With no probability information being input, the output can only be view as discrete realizations of the response. The epistemic interval is then determined from the maximum and minimum values to the response, can be viewed as the extreme or "worst case" analysis.

An additional, special case of epistemic uncertainty is numerical error. This uncertainty is common in numerical modelling and is defined as a recognizable deficiency in any phase or activity of modelling and simulations that is not due to lack of knowledge of the physical system. In CFD, an example of this type of uncertainty would be the discretization error in both the temporal and spatial domains that comes from the numerical solution of the partial differential equations that govern the system. This uncertainty can be well understood and controlled through code verification and grid convergence studies.

3. Demonstration on an Analytical Function

Before investigating a large-scale planetary entry problem, the sparse approximation approach is first applied to a simple polynomial function, which is shown in Eq.(3.16). This function can quickly evaluated with a Monte Carlo analysis to verify the results when using a sparse approximation of the PCE.

$$F = \sum_{i=0}^{1}(x_{i+i} - x_i^2)^2 + (1 - x_i)^2 + \sum_{j=2}^{8} x_{j+1}. \tag{3.16}$$

The first term in this equation is the classic Rosenbrock function, which is a fourth order polynomial often used in optimization testing. The second term is just a series of linear terms that are added to the output of the Rosenbrock function. This can be viewed as noise added to the measurement. In total, there are 10 random variables in this function. Let each of the variables, x, be on a uniform distribution on the interval [0.5 , 1.5]. Note that on this domain, the linear terms added to the Rosenbrock function will be insignificant compared to the other three variables. This is done intentionally to represent a system where only a small fraction of the random variables are important.

Following the approach in Sec. 2.3, an initial sample structure is constructed containing 1001 samples, which is the number of terms in a forth order PCE with 10 random variables (see Eq. (3.3)). Then, 5 samples were iteratively pulled from this sample set to solve the optimization problem in Eq. (3.5) with an increasing number of samples, up to 1000. From Sec. 2.3, the convergence of the mean Sobol index error is shown in Fig. 3.1(a).

The error drops rapidly reaching a value near 10^{-10} with less than 100 samples. This demonstrates that the optimization approach was able to quickly recover a sparse approximation of the PCE with a sample size much less than that required for a direct solution of the linear system in Eq. (3.4).

For validation purposes, the uncertainty in the model can be propagated through the surrogate model at each sample size iteration and compared to the results obtained from a Monte Carlo analysis performed on the deterministic model. The convergence of the 95% confidence interval is shown in Fig. 3.1(b), compared to a Monte Carlo analysis with 10^6 samples. Note that because all of the uncertainty is aleatory, a cumulative density function can be constructed with all of the outputs from sampling the surrogate model. Statistics, such as confidence intervals can extracted from the outputs.

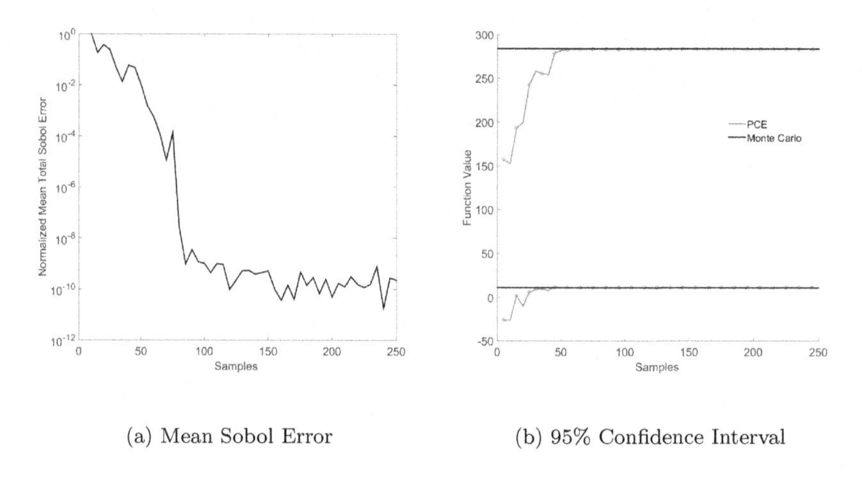

(a) Mean Sobol Error

(b) 95% Confidence Interval

Fig. 3.1. Analytical function convergence.

4. Prediction of Radiative Heating for Mars Entry

To demonstrate the implementation of the sparse approximation approach on a large-scale engineering problem, the uncertainty analysis of radiative heating for Mars entry is investigated. This particular problem is ideal for demonstrating the key aspects of the uncertainty quantification approach as it deals with a high-fidelity, expensive CFD simulation that contains a significant amount of uncertainty originating from various sources associated with the physical modelling.

Note that in this section, significant detail is provided regarding the computational model to make the problem repeatable for the interested reader. In the results, focus is given to how the sparse approximation approach is used and evaluated. The reader is strongly encouraged to survey the provided references for a more detailed discussion of the physical phenomena that govern the radiation. The remainder of this section details deterministic model, the stochastic problem, and the results of the uncertainty analysis.

4.1. *Geometry, flow solver, and entry conditions*

In this study, the flow field was modelled using the Langley Aerothermodynamic Upwind Relaxation Algorithm (LAURA) finite-volume, Navier-Stokes flow solver.[16] This solver uses a second-order, upwind, discretization scheme with relaxation of both inviscid and viscous terms for solution stability. LAURA has been used for many high energy flow studies and has been extensively validated for various atmospheric entry flow scenarios. For both entry cases in this study, the flow field was assumed to be at steady state and was modelled using a two-temperature, thermochemical nonequilibrium model.[17,18] A constant 1500 K, super-catalytic wall boundary condition is assumed, the same for each case. Note that the super-catalytic assumption has a negligible effect on the radiative heating.[19]

The radiation was modelled using the High Temperature Aerothermodynamic Radiation (HARA) code.[20,21] This nonequilibrium radiation code uses a tangent-slab approximation for computing the radiative flux and its divergence. HARA is based on a set of atomic levels and lines obtained from the National Institute of Standards and Technology (NIST) database,[22] Opacity Project databases,[23] and atomic bound-free (photoionization) cross-sections from the TOPbase.[24] A total of 16 molecular band systems are modelled for this Mars entry scenario. The oscillator strengths applied to the C_2, CN, CO 4th Positive, and CO Infrared band systems are presented by Babou *et al.*[25] da Sylva and Dudeck[26] detail the oscillator strengths for the remaining CO band systems. HARA uses a Collisional Radiative (CR) or non-Boltzmann modelling of atomic and molecular electronic states. This is based on a set of electronic and heavy particle impact excitation rates. The non-Boltzmann approach and rate models are presented in detail by Johnston *et al.*[19] Note that the flow field solver and the radiative heat transfer calculations are coupled as coupling can significantly affect the radiation prediction.[27]

The geometry of interest in this study is a Hypersonic Inflatable Aerodynamic Decelerator (HIAD) fore-body, which is illustrated in Fig. 3.2(a). The HIAD heat shield geometry is modelled as a 70 degree spherical cone with a nose radius of 3.75 m, a shoulder radius of 0.375 m, and a base radius of 7.5 m. The computational grid used for this geometry is 128×48, which has been shown to be sufficient to capture the radiation.[19,28] The grid varies based on the shock location. LAURA uses a gradient-based shock capturing technique to detect and cluster the grid in the flow direction. A sample of the grid is shown in Fig. 3.2(b).

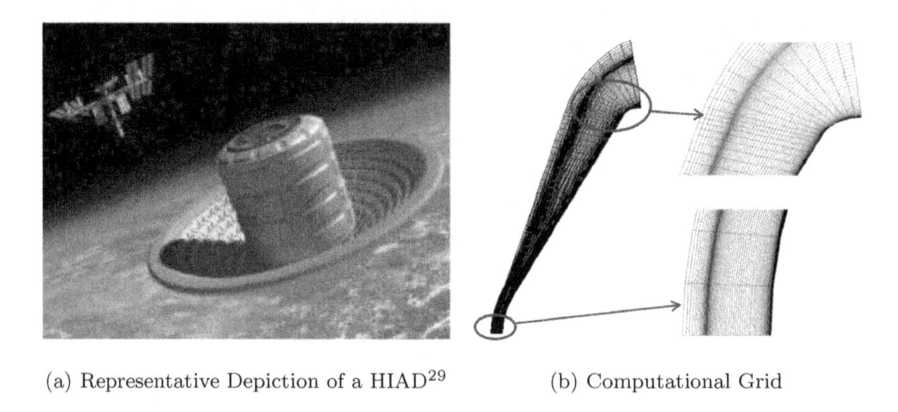

(a) Representative Depiction of a HIAD[29] (b) Computational Grid

Fig. 3.2. Hypersonic inflatable aerodynamic decelerator.

The Martian atmospheric composition was assumed to be composed of 96% CO_2 and 4% N_2, per mole. The flow field was modelled with a 17 species composition model: CO_2, CO, N_2, O_2, NO, C, N, O, CN, C_2, C^+, O^+, CO^+, O_2^+, NO^+, CN^+, and e^-. The chemical reaction model, presented by Johnston *et al.*,[19] is composed of 34 reactions that includes dissociation and exchange mechanisms, as well as ionization reactions. The freestream velocity, density, and temperature conditions for the Mars entry were selected to be 7.0 km/s, 1.0e−4 kg/m^3, and 150 K, respectively. These values are typical of super-orbital entry velocities into Mars, similar to the entry conditions observed in the Mars Pathfinder mission.

4.1.1. *Baseline solution*

When performing the UQ analysis, a converged, baseline solution is used as the starting point for the CFD evaluation at each sample location to accelerate the solution convergence. Temperature and pressure contours of the flow field are shown in Figs. 3.3(a) and 3.3(b), respectively. Additionally, Fig. 3.3(c) shows that shock stand-off distance and Fig. 3.3(d) shows the wall radiative heat flux distribution along the surface of the HIAD. These figures predominately show the shock is well defined in the flow field and grows linearly from the flat portion of the vehicle. A significant portion of the radiation emitted from the shock layer comes from a strong nonequilibrium region, just behind the shock. A well resolved shock region is required to accurately capture the flow field properties and chemical composition near the shock.

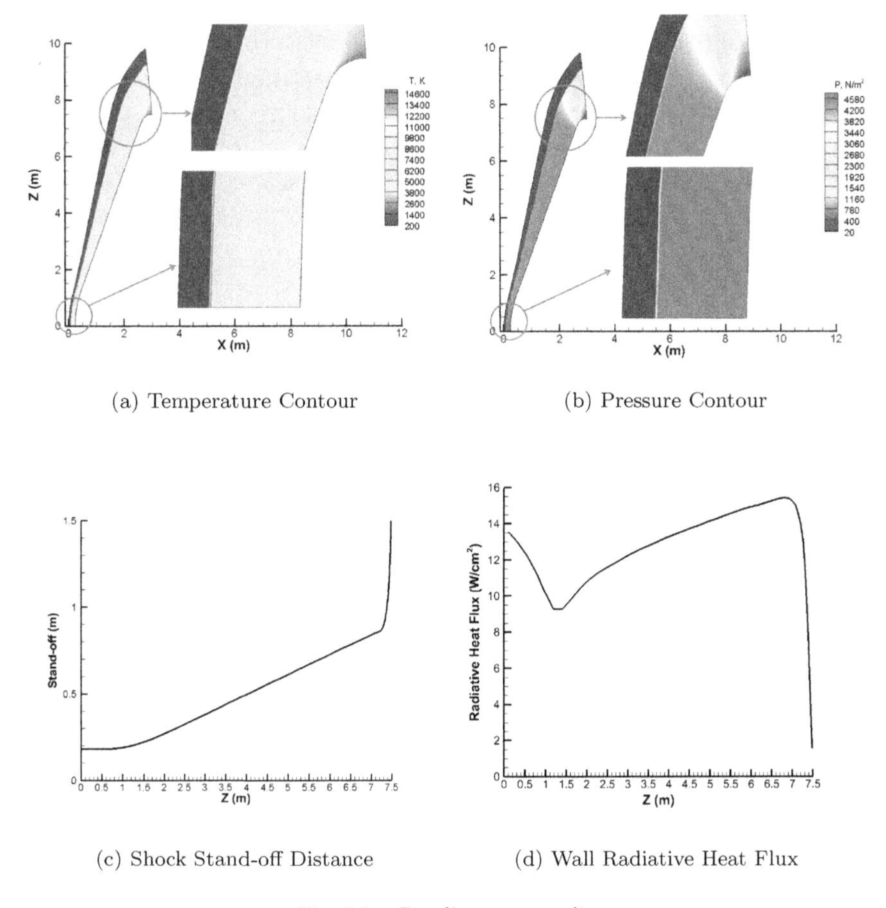

(a) Temperature Contour

(b) Pressure Contour

(c) Shock Stand-off Distance

(d) Wall Radiative Heat Flux

Fig. 3.3. Baseline case results.

Temperature and heat flux distributions along the stagnation line are shown in Figs. 3.4(a) and 3.4(b), respectively. The nonequilibrium cases are compared to the equilibrium solution to highlight the significance of nonequilibrium. At the stagnation point, notice the region of strong nonequilibrium at the shock location. The translational temperature spikes to nearly 16,000 K while the vibrational temperature peaks just below 10,000 K. Shortly after the shock, however, the two temperatures coalesce, indicating the presence of thermal equilibrium. At the stagnation point, the radiation at the surface is primarily due to the emission from the CO 4th Positive band. This is shown in the radiative flux spectrum given in Fig. 3.4(c) by the large region of emission from the CO 4th Positive band wavelength. The decrease in the radiative flux through the shock layer is due to the optical thickness of the CO 4th Positive band system, which emits in the vacuum ultraviolet region of the spectrum, and therefore experiences significant self-absorption for most shock layer conditions. Notice also in this figure that there is significant contribution from the CN Violet band, which emits primarily from the equilibrium region of the shock layer.

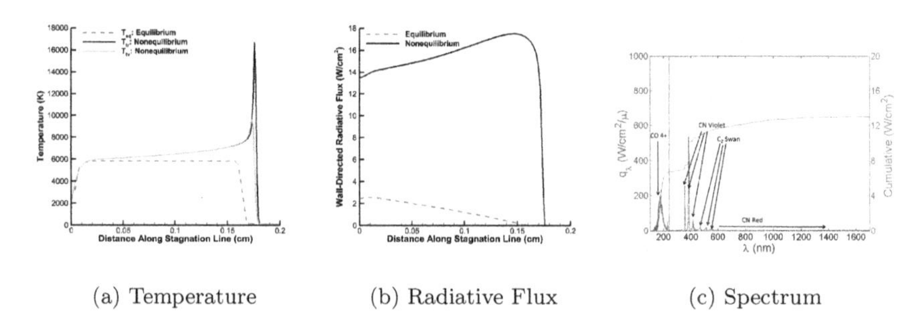

(a) Temperature (b) Radiative Flux (c) Spectrum

Fig. 3.4. Mars baseline stagnation line temperature and radiative heat flux distributions.

Moving along the HIAD surface, the nonequilibrium region of the shock layer begins to decrease as the difference between the two temperatures at the shock goes down and the temperature within the shock layer approaches the equilibrium temperature. Figure 3.5(a) shows that the difference between the translation and vibrational temperatures is less at the stagnation region. This indicates a reduction in the thermal nonequilibrium region of the shock. Referring back to Fig. 3.3(d), there is an initial decrease in the radiative heat flux. The volumetric radiance from the CO 4th Positive band system drops by nearly a factor of two between the stagnation point and the sphere-cone juncture, which causes the reduced radiation at the surface. This is reflected in Fig. 3.5(b) at $Z = 1.81$ m normal to the stagnation line. The spectrum in Fig. 3.5(c) also shows the reduction in the CO 4th Positive emission.

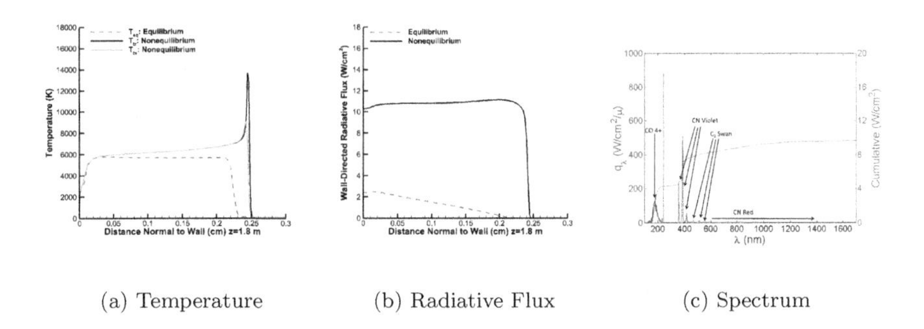

(a) Temperature (b) Radiative Flux (c) Spectrum

Fig. 3.5. Mars baseline temperature and radiative heat flux distributions at $Z = 1.81$ m.

The reduction in the nonequilibrium region of the shock layer decreases further, moving towards the shoulder region of the HIAD. Figure 3.6(a) at $Z = 6.69$ meters normal to the stagnation line, near the shoulder shows that the flow is in nearly thermochemical equilibrium and that the radiation is increasing through the shock layer, as seen in Fig. 3.6(b). Referring back to Fig. 3.3(d), the radiative flux at the wall is actually increasing from the sphere-cone juncture to the shoulder. This has two reasons. The first is the increase in the shock stand-off distance. Radiation from the shock layer, in the simplest case, is known to be linearly dependent on shock stand-off distance. The second reason is related to a change in the emission spectrum. As the flow nears equilibrium, emission by the CN Violet band system becomes significant. This is confirmed by looking at the spectrum in Fig. 3.6(c), which shows the peak of the CN violet states nearly double that of the stagnation region levels. This band is also optically thin and the CN molecule forms in equilibrium concentrations through the shock layer. With the increasing shock stand-off distance along the surface of the HIAD, the CN Violet band has an increasing optical path to emit across, causing an increase in the wall-directed radiative flux. Also note that, while not shown here, there is a small contribution from the CO Infrared band system coming from the 4800–5200 nm wavelength range. This is due to the deexcitation of the CO molecule to its ground state caused by the lower temperatures at the shoulder region.

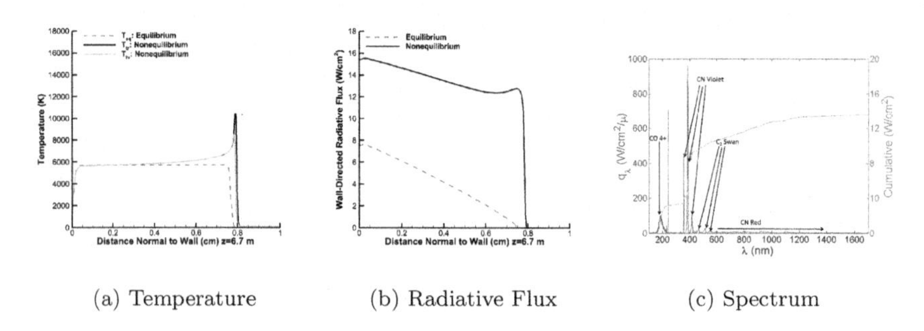

(a) Temperature (b) Radiative Flux (c) Spectrum

Fig. 3.6. Mars baseline temperature and radiative heat flux distributions at $Z = 6.69$ m.

4.1.2. *Uncertain parameters*

The uncertain parameters in this problem come from two primary sources: the flow field chemistry and the radiation modelling. A 34-reaction model is used to simulate the flow field chemistry. Twenty of these reactions, listed in Table 3.1, are dissociation and exchange reactions and are considered to be uncertain due to the variation in the reaction rates. Uncertainty in the remaining 14 ionization rates has been shown to have an insignificant effect on the radiation under these entry conditions and are, therefore, neglected from the uncertainty analysis, but are considered as part of the finite rate chemistry model. The details of the baseline reaction rates are provided in the work by Johnston *et al.*[19]

The uncertainty in the radiation modelling has been identified to come from two sources. The first is the oscillator strengths of the molecular band systems. In this problem, 16 band systems are treated. The results in the previous section showed that the CO, CN, C_2 contribute significantly to the radiation. A list of the treated band systems and their uncertainty are given in Table 3.2. Notice that, while not significant radiation contributors at these conditions, two CO_2 systems are treated as CO_2 is in abundance in the Martian atmosphere and does have the potential to radiate.

The second group of uncertain parameters in the radiation modelling is due to the uncertainty in the non-Boltzmann modelling of excited states. Similar to the molecular band systems, the electronic states of CO, CN, and C_2 are considered uncertain as these band systems most influence the radiation. In total, there are 22 heavy particle impact excitation rates and 35 electron impact excitation rates considered uncertain. These reaction mechanisms are shown in Tables 3.3 and 3.4, respectively. Details of each individual mechanism are provided in the work by Johnston *et al.*[19]

Table 3.1. Mars flow field chemical kinetics.

#	Reaction	Uncertainty
1	$CO_2 + M \leftrightarrow CO + O + M$	−1, +0 om
		−1, +0 om
2	$CO + M \leftrightarrow C + O + M$	−75%, +50%
3	$C_2 + M \leftrightarrow 2C + M$	−1, +1 om
4	$CN + M \leftrightarrow C + N + M$	−1, +1 om
5	$N_2 + M \leftrightarrow 2N + M$	−1, +1 om
		−1, +1 om
		−1, +1 om
6	$NO + M \leftrightarrow N + O + M$	−1, +1 om
		−1, +1 om
7	$O2 + M \leftrightarrow 2O + M$	−50%, +50%
		−50%, +50%
8	$CO_2 + O \leftrightarrow O_2 + CO$	−1, +1 om
9	$CO + C \leftrightarrow C_2 + O$	−1, +1 om
10	$CO + N \leftrightarrow CN + O$	−1, +1 om
11	$CO + NO \leftrightarrow CO_2 + N$	−1, +1 om
12	$CO + O \leftrightarrow O_2 + C$	−0, +1 om
13	$C_2 + N_2 \leftrightarrow CN + CN$	−1, +1 om
14	$CN + C \leftrightarrow C_2 + N$	−1, +1 om
15	$CN + O \leftrightarrow NO + C$	−1, +1 om
16	$N + CO \leftrightarrow NO + C$	−1, +1 om
17	$N_2 + C \leftrightarrow CN + N$	−50%, +50%
18	$N_2 + CO \leftrightarrow CN + NO$	−1, +1 om
19	$N_2 + O \leftrightarrow NO + N$	−50%, +50%
20	$O_2 + N \leftrightarrow NO + O$	−1, +1 om

4.2. *Uncertainty Analysis*

To investigate the effect of uncertainty classification, two uncertainty analyses are performed: a pure epistemic analysis and a pure aleatory analysis. All of the uncertain parameters that will be considered are bounded with specified limits, meaning that the same surrogate models created with the Legendre basis functions could be used for both epistemic and the aleatory analysis. An important note is that while the aleatory uncertainty analysis considers uniform probability for each uncertain variable in their specified intervals, no probability assignment is made in the epistemic analysis. The purpose of performing these two analyses is to show the effect of uncertain parameter classification, interpretation of output uncertainty, and the importance of correctly representing the uncertain parameters.

A total number of 93 uncertain parameters are considered in the uncertainty analysis. A second order PCE for this case will have 4,465 terms, calculated using Eq. (3.3). This corresponds to the minimum number of CFD model evaluations that would be necessary to directly solve the linear

Table 3.2. Molecular band processes.

Molecule	Band Name	λ Range (nm)	Uncertainty
CO	4th Positive	120 – 280	+/– 40%
CO	3rd Positive	250 – 450	+/– 50%
CO	Triplet	320 – 2500	+/– 50%
CO	Asundi	370 – 2500	+/– 50%
CO	Angstrom	400 – 700	+/– 50%
CO	Infrared	1200 – 7000	+/– 50%
CN	Red	400 – 2800	+/– 30%
CN	Violet	300 – 550	+/– 15%
C_2	Swan	390 – 1000	+/– 50%
C_2	Ballik-Ramsay	500 – 3000	+/– 50%
C_2	Phillips	350 – 1200	+/– 50%
C_2	Mulliken	200 – 250	+/– 50%
C_2	Des.-D'Azam.	280 – 700	+/– 50%
C_2	Fox-Herzberg	200 – 500	+/– 50%
CO_2	Infrared	1700 – 25000	+/– 50%
CO_2	UV	190 – 320	+/– 100%

system in Eq. (3.4) for the PCE coefficients. To reduce this computational expense, the sparse approximation technique can be employed. The optimization routine in Eq. (3.5) was solved with an iteratively increased sample set, from 10 samples to 500 samples, increased by 10 samples at each iteration. Convergence of the Sobol indices at three locations along the surface of the HIAD are shown in Fig. 3.7(a). In 500 samples, the error drops to about 2% at three points along the surface of the HIAD and is an indication of the convergence of the PCE coefficients.

To determine the accuracy of the surrogate model, 100 new test points were generated in the design space and the actual, deterministic model values were compared to the values from surrogate response surface at these locations. A plot of the normalized test point error is shown in Fig. 3.7(b). With 500 samples, the test point error is about 10%. Given the nature and complexity of the CFD model problem, this error is not unexpected and has been observed in other studies using different techniques. Both errors are deemed acceptable given the near 90% drop in computational cost; however, this error may be further reduced with additional samples.

Surrogate models were constructed at 13, evenly spaced points along the surface of the HIAD for the radiative heat flux. The epistemic and the 95% confidence intervals of the wall radiative heat flux is shown in Fig. 3.8. Notice that there is significant variation from the baseline value. After the sphere-cone juncture, the uncertainty intervals remain relatively constant. The uncertainty intervals are shown to be the largest at the stagnation

Table 3.3. uncertain heavy-particle impact excitation reactions.

#	Reaction	Uncertainty
1	$CN(X^2\Sigma^+) + M \leftrightarrow CN(A^2\Pi) + M$	+/− 1 om
2	$CN(A^2\Pi\ ) + M \leftrightarrow CN(B^2\Sigma^+) + M$	+/− 1 om
3	$CN(B^2\Sigma^+\ ) + M \leftrightarrow CN(a^4\Sigma^+) + M$	+/− 2 om
4	$CN(a^4\Sigma^+) + M \leftrightarrow CN(D^2\Pi^+) + M$	+/− 2 om
5	$CO(X^1\Sigma^+) + M \leftrightarrow CO(a^3\Pi) + M$	+/− 1 om
6	$CO(X^1\Sigma^+) + M \leftrightarrow CO(a'^3\Sigma^+) + M$	+/− 1 om
7	$CO(X^1\Sigma^+) + M \leftrightarrow CO(d^3\Delta) + M$	+/− 1 om
8	$CO(X^1\Sigma^+) + M \leftrightarrow CO(A^1\Pi) + M$	+/− 1 om
9	$CO(a^3\Pi) + M \leftrightarrow CO(a'^3\Sigma^+) + M$	+/− 2 om
10	$CO(a'^3\Sigma^+) + M \leftrightarrow CO(d^3\Delta) + M$	+/− 2 om
11	$CO(d^3\Delta) + M \leftrightarrow CO(e^3\Sigma^-) + M$	+/− 2 om
12	$CO(e^3\Sigma^-) + M \leftrightarrow CO(A^1\Pi) + M$	+/− 2 om
13	$C_2(X^1\Sigma^+) + M \leftrightarrow C_2(b^3\Sigma^-) + M$	+/− 2 om
14	$C_2(X^1\Sigma^+) + M \leftrightarrow C_2(c^3\Sigma^+) + M$	+/− 2 om
15	$C_2(X^1\Sigma^+) + M \leftrightarrow C_2(d^3\Pi) + M$	+/− 1 om
16	$C_2(X^1\Sigma^+) + M \leftrightarrow C_2(C^1\Pi) + M$	+/− 2 om
17	$C_2(b^3\Sigma^-) + M \leftrightarrow C_2(c^3\Sigma^+) + M$	+/− 2 om
18	$C_2(b^3\Sigma^-) + M \leftrightarrow C_2(d^3\Pi) + M$	+/− 2 om
19	$C_2(b^3\Sigma^-) + M \leftrightarrow C_2(C^1\Pi) + M$	+/− 2 om
20	$C_2(c^3\Sigma^+) + M \leftrightarrow C_2(d^3\Pi) + M$	+/− 2 om
21	$C_2(c^3\Sigma^+) + M \leftrightarrow C_2(C^1\Pi) + M$	+/− 2 om
22	$C_2(d^3\Pi) + M \leftrightarrow C_2(C^1\Pi) + M$	+/− 2 om

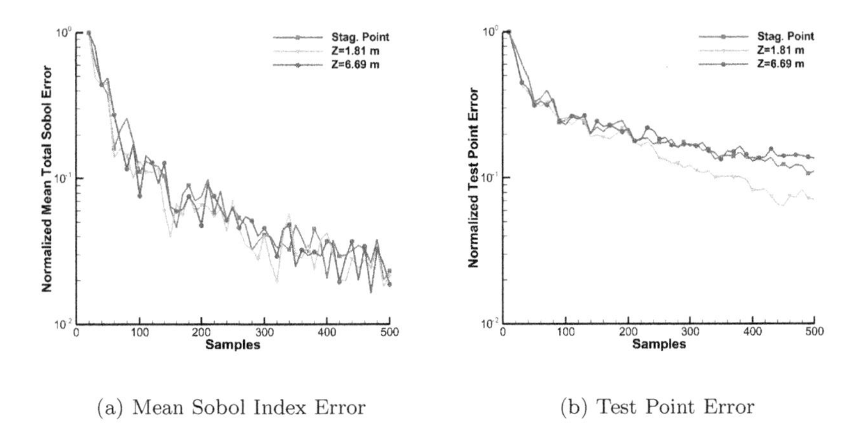

(a) Mean Sobol Index Error

(b) Test Point Error

Fig. 3.7. Convergence of sparse PCEs for Mars entry.

point, indicating that the uncertainty in the inputs have the most influence in this region.

Also from Fig. 3.8, the importance of correct characterization of the

Table 3.4. Uncertain electron impact excitation reactions.

#	Reaction	Uncertainty
1	$CN(X^2\Sigma^+) + e^- \leftrightarrow CN(A^2\Pi) + e^-$	+/− 1 om
2	$CN(X^2\Sigma^+) + e^- \leftrightarrow CN(B^2\Sigma^+) + e^-$	+/− 1 om
3	$CN(X^2\Sigma^+) + e^- \leftrightarrow CN(a^4\Sigma^+) + e^-$	+/− 2 om
4	$CN(X^2\Sigma^+) + e^- \leftrightarrow CN(D^2\Pi^+) + e^-$	+/− 2 om
5	$CN(A^2\Pi) + e^- \leftrightarrow CN(B^2\Sigma^+) + e^-$	+/− 2 om
6	$CN(A^2\Pi) + e^- \leftrightarrow CN(a^4\Sigma^+) + e^-$	+/− 2 om
7	$CN(A^2\Pi) + e^- \leftrightarrow CN(D^2\Pi^+) + e^-$	+/− 2 om
8	$CN(B^2\Sigma^+) + e^- \leftrightarrow CN(a^4\Sigma^+) + e^-$	+/− 2 om
9	$CN(B^2\Sigma^+) + e^- \leftrightarrow CN(D^2\Pi^+) + e^-$	+/− 2 om
10	$CN(a^4\Sigma^+) + e^- \leftrightarrow CN(D^2\Pi^+) + e^-$	+/− 2 om
11	$CO(X^1\Sigma^+) + e^- \leftrightarrow CO(a^3\Pi)+ e^-$	+/− 1 om
12	$CO(X^1\Sigma^+) + e^- \leftrightarrow CO(a'^3\Sigma^+)+ e^-$	+/− 1 om
13	$CO(X^1\Sigma^+) + e^- \leftrightarrow CO(d^3\Delta)+ e^-$	+/− 1 om
14	$CO(X^1\Sigma^+) + e^- \leftrightarrow CO(e^3\Sigma^-)+ e^-$	+/− 1 om
15	$CO(X^1\Sigma^+) + e^- \leftrightarrow CO(A^1\Pi)+ e^-$	+/− 1 om
16	$CO(a^3\Pi) + e^- \leftrightarrow CO(a'^3\Sigma^+)+ e^-$	+/− 1 om
17	$CO(a^3\Pi) + e^- \leftrightarrow CO(d^3\Delta)+ e^-$	+/− 1 om
18	$CO(a^3\Pi) + e^- \leftrightarrow CO(e^3\Sigma^-)+ e^-$	+/− 1 om
19	$CO(a^3\Pi) + e^- \leftrightarrow CO(A^1\Pi)+ e^-$	+/− 2 om
20	$CO(a'^3\Sigma^+) + e^- \leftrightarrow CO(d^3\Delta)+ e^-$	+/− 1 om
21	$CO(a'^3\Sigma^+) + e^- \leftrightarrow CO(e^3\Sigma^-)+ e^-$	+/− 1 om
22	$CO(a'^3\Sigma^+) + e^- \leftrightarrow CO(A^1\Pi)+ e^-$	+/− 2 om
23	$CO(d^3\Delta) + e^- \leftrightarrow CO(e^3\Sigma^-)+ e^-$	+/− 1 om
24	$CO(d^3\Delta) + e^- \leftrightarrow CO(A^1\Pi)+ e^-$	+/− 2 om
25	$CO(e^3\Sigma^-) + e^- \leftrightarrow CO(A^1\Pi)+ e^-$	+/− 2 om
26	$C_2(X^1\Sigma^+) + e^- \leftrightarrow C_2(b^3\Sigma^-) + e^-$	+/− 2 om
27	$C_2(X^1\Sigma^+) + e^- \leftrightarrow C_2(c^3\Sigma^+) + e^-$	+/− 2 om
28	$C_2(X^1\Sigma^+) + e^- \leftrightarrow C_2(d^3\Pi) + e^-$	+/− 1 om
29	$C_2(X^1\Sigma^+) + e^- \leftrightarrow C_2(C^1\Pi) + e^-$	+/− 2 om
30	$C_2(b^3\Sigma^-) + e^- \leftrightarrow C_2(c^3\Sigma^+) + e^-$	+/− 2 om
31	$C_2(b^3\Sigma^-) + e^- \leftrightarrow C_2(d^3\Pi) + e^-$	+/− 2 om
32	$C_2(b^3\Sigma^-) + e^- \leftrightarrow C_2(C^1\Pi) + e^-$	+/− 2 om
33	$C_2(c^3\Sigma^+) + e^- \leftrightarrow C_2(d^3\Pi) + e^-$	+/− 2 om
34	$C_2(c^3\Sigma^+) + e^- \leftrightarrow C_2(C^1\Pi) + e^-$	+/− 2 om
35	$C_2(d^3\Pi) + e^- \leftrightarrow C_2(C^1\Pi) + e^-$	+/− 2 om

uncertainty is illustrated as there is a significant difference between the non-probabilistic, epistemic interval and the probabilistic, aleatory interval. The most appropriate analysis may be a mixed (epistemic and aleatory) analysis. However, given the nature of the uncertain parameters in this problem and in many other engineering applications, this may not be possible as assigning probabilistic distributions to the inputs considered in the present study may be difficult due to the lack of knowledge of their behaviour.

Further confirmation of the convergence of the PCE coefficients can

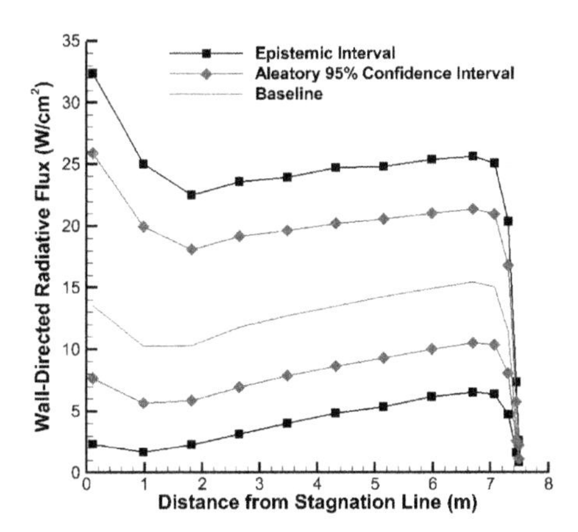

Fig. 3.8. Epistemic and 95% confidence intervals of wall radiative heat flux for Mars entry.

be done by monitoring the change in the epistemic and 95% confidence intervals at selected locations on the HIAD surface. At the stagnation point, convergence of both intervals are shown in Fig. 3.9(a). Additionally, convergence of the intervals at $Z = 1.81$ m and $Z = 6.69$ m normal to the stagnation line are shown in Figs. 3.9(b) and 3.9(c), respectively. These figures clearly show the convergence of the PCEs, as well as the effect of the input uncertainty. Actually, these figures suggest that convergence of the PCEs are occurring much faster than what was shown previously in Fig. 3.7. After about 100 to 200 samples, only slight changes occur in the intervals. Depending on the desired outcome from the analysis, different quantities may be tracked for convergence. While the convergence of the output intervals may achieved more rapidly (for this problem) a more accurate surrogate model and variable ranking may be achieved by further converging the PCE coefficients.

4.3. *Sensitivity analysis*

Accurately knowing the output variance of the radiative heat flux is critical in ensuring reliable and robust design of a thermal protection system. Reducing the output variation may be achieved by reducing the uncertainty of the input parameters by improving the knowledge associated with the uncertain parameters. The most efficient approach is to focus on the input

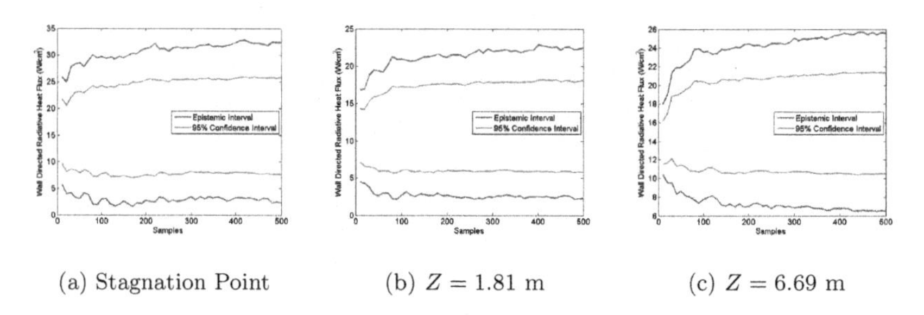

(a) Stagnation Point (b) $Z = 1.81$ m (c) $Z = 6.69$ m

Fig. 3.9. Convergence of the radiative heating uncertainty intervals for Mars entry.

parameters that contribute most significantly to the total output uncertainty. This requires a sensitivity analysis of the radiative heat flux to the input parameters. Because the Sobol index of each uncertain parameter was used to measure the convergence of the PCE, the global, nonlinear sensitivities are already determined. Contributions from the four groups of uncertain parameters are shown in Figs. 3.10(a)–3.10(c), at the stagnation point, $Z = 1.81$ m, and $Z = 6.69$ m, respectively.

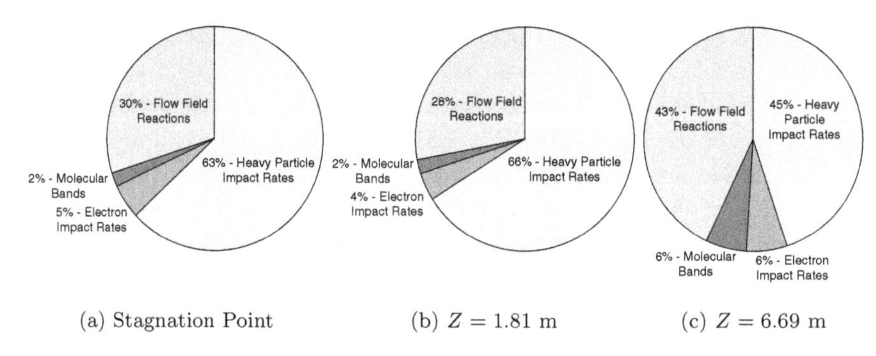

(a) Stagnation Point (b) $Z = 1.81$ m (c) $Z = 6.69$ m

Fig. 3.10. Uncertainty contributions to Mars radiative heating along the HIAD surface.

Notice first that the radiative heating uncertainty at the stagnation point and at the sphere-cone juncture is dominated by the uncertainty in the heavy particle impact excitation rates. The next main contributor is the uncertainty in the flow field reactions. Note that the contributions from the molecular band and electron impact excitation rates are small in comparison to the other two groups. Moving towards the shoulder, there is a decrease in the heavy particle impact contribution and an increase in the flow field reaction contribution. As discussed above, identifying the significant sources of uncertainty can be used to assist in resource allocation for improving understanding of the uncertain variables. For example, experiments may be dedicated to improving the understanding of the heavy particle impact excitation rates.

5. Summary

In this chapter, the objective was to outline an approach to uncertainty quantification for complex computation models with a large number of uncertain parameters, such as those models used for the prediction of radiative heat flux during hypersonic, planetary entry. The challenge was that computational models of these flows are computationally expensive and possess a significant amount of uncertainty, which makes a Monte Carlo analysis or the use of many surrogate modelling approaches impractical. To overcome this, an approach for the sparse approximation of a polynomial chaos expansion (PCE) was introduced as a means of efficient uncertainty quantification. The fundamental idea was to use Basis Pursuit Denoising to recover the PCE coefficients from an underdetermined system of linear equations. For improved computational efficiency, solutions were obtained iteratively with increasing sample size while tracking the convergence of the PCE coefficients. Two methods were introduced to measure the accuracy of the expansion coefficients and their convergence when the sample size used to obtain the PCE coefficients was iteratively increased. These two methods included using both the sensitivities of each uncertain parameter via the calculation of Sobol indices and a comparison to the actual response at selected test points in the design space.

The sparse approximation technique was used to investigate the uncertainty in radiative heating of a hypersonic inflatable aerodynamics decelerator during Mars entry. The iterative sampling procedure indicated that accurate results were obtained with about 500 samples, which is only about 10% of the number needed for an exact solution to the linear system obtained with the Point-Collocation Non-intrusive Polynomial Chaos method. Epistemic and aleatory uncertainty analyses were performed to highlight the importance of uncertain parameter classification. The uncertainty analysis showed that the there is significant variability from the baseline solution due to the presence of the input uncertainty. A sensitivity analysis highlighted those parameters that contribute most significantly to the radiation uncertainty.

Overall the technique presented here is a powerful tool for conducting an uncertainty analysis of computationally demanding models with a large number of uncertain variables. Analysis of problems of this type has long been a significant challenge, and this approach makes progress in allowing for the analysis of such challenging engineering models.

References

1. S. Hosder and B. Bettis, Uncertainty and sensitivity analysis for reentry flows with inherent and model-form uncertainties, *J. Spacecr. Rockets* **49**(2), 193–206 (2012).
2. B. Bettis, S. Hosder and T. Winter, Efficient uncertainty quantification in multidisciplinary analysis of a reusable launch vehicle, AIAA Paper 2011-2393 (2011).
3. S. Hosder, R. W. Walters and M. Balch, Point-collocation nonintrusive polynomial chaos method for stochastic computational fluid dynamics, *AIAA J.* **48**(12), 2721–2730 (2010).
4. J. A. S. Witteveen and H. Bijl, Efficient quantification of the effect of uncertainties in advection–diffusion problems using polynomial chaos, *Numer. Heat Transf.* **53**(5), 437–465 (2008).
5. D. Han and S. Hosder, Inherent and model-form uncertainty analysis for CFD simulation of synthetic jet actuators, AIAA Paper 2012-0082 (2012).
6. R. G. Ghanem and P. D. Spanos, *Stochastic Finite Elements: A Spectral Approach.* Springer-Verlag, New York (1991).
7. D. Xiu and G. M. Karniadakis, The Wiener-Askey polynomial chaos for stochastic differential equations, *SIAM J. Sci. Comput.* **24**(2), 619–644 (2002).
8. M. S. Eldred, Recent advances in non-intrusive polynomial chaos and stochastic collocation methods for uncertainty analysis and design, AIAA Paper 2009-2274 (2009).
9. T. K. West IV, S. Hosder and C. O. Johnston, Multi-step uncertainty quantification approach applied to hypersonic reentry flows, *J. Spacecr. Rockets* **51**(1), 296–310 (2014).
10. S. Hosder, R. W. Walters and M. Balch, Efficient sampling for non-intrusive polynomial chaos applications with multiple uncertain input variables, AIAA Paper 2007-1939 (2007).
11. A. Doostan and H. Owhadi, A non-adapted sparse approximation of PDES with stochastic inputs, *J. Comput. Phys.* **230**(8), 3015–3034 (2011).
12. A. Yang, A. Ganesh, S. Sastry and Y. Ma, Fast l1-minimization algorithms and an application in robust face recognition: A review, Technical Report UCB/EECS-2010-13, EECS Department, University of California, Berkeley (2010).
13. M. S. Asif and J. Romberg, Fast and accurate algorithms for re-weighted l_1-norm minimization, *IEEE Trans. Signal Process.* **61**(23), 5905–4916 (2013).
14. B. Sudret, Global sensitivity analysis using polynomial chaos expansion, *Reliab. Eng. Syst. Saf.* **93**(7), 964–979 (2008).
15. W. L. Oberkampf, J. C. Helton and K. Sentz, Mathematical representation of uncertainty, AIAA Paper 2001-1645 (2001).
16. A. Mazaheri, P. A. Gnoffo, C. O. Johnston and B. Kleb, Laura users manual: 5.4-54166, Technical Report, NASA/TM-2011-217092 (2009).

17. P. A. Gnoffo, R. N. Gupta and J. L. Shinn, Conservation equations and physical models for hypersonic air flows in thermal and chemical nonequilibrium, Technical report, NASA TP 2867 (1989).
18. C. Park, J. T. Howe, R. L. Jaffe and G. V. Candle, Review of chemical-kinetic problems for future NASA missions, II: Mars entries, *J. Thermophys. Heat Transf.* **8**(1), 9–23 (1994).
19. C. O. Johnston, A. M. Brandis and K. Sutton, Shock layer radiation modeling and uncertainty for mars entry, AIAA Paper 2012-2866 (2012).
20. C. O. Johnson, B. R. Hollis and K. Sutton, Spectrum modeling for air shock-layer radiation at lunar-return conditions, *J. Spacecr. Rockets* **45**(5), 865–878 (2008).
21. C. O. Johnson, B. R. Hollis and K. Sutton, Non-Boltzmann modeling for air shock layers at lunar return conditions, *J. Spacecr. Rockets* **45**(5), 879–890 (2008).
22. Y. Ralchenko, NIST atomic spectra database, version 3.1.0 URL `physics.nist.gov/PhysRefData/ASD/`.
23. Opacity Project Team, *The Opacity Project*, Vol. 1, Institute of Physics Publishing, London, 1995.
24. W. Cunto, C. Mendoza, F. Ochsenbein and C. Zeippen, Topbase at the CDS, *Astron. Astrophys.* **275**, L5–L8 (1993).
25. Y. Babou, P. Riviere, M. Perrin and A. Soufiani, Spectroscopic data for the prediction of radiative transfer in Co_2–N_2 plasmas, *J. Quant. Spectrosc. Radiat. Transf.* **110**, 89–108 (2009).
26. M. L. da Silva and M. Dudeck, Arrays of radiative transition probabilities for Co_2–N_2 plasmas, *J. Quant. Spectrosc. Radiat. Transf.* **102**, 348–386 (2006).
27. M. Wright, D. Bose and J. Olejniczak, Impact of flowfield-radiation coupling on aeroheating for titan aerocapture, *J. Thermophys. Heat Transf.* **19**(1), 17–27 (2005).
28. T. K. West IV, A. J. Brune, S. Hosder and C. O. Johnston, Uncertainty analysis of radiative heating predictions for titan entry, *J. Thermophys. Heat Transf.* (2015), doi: 102514/1.T4620.
29. www.nasa.gov.

Chapter 4

Investigation About Uncertain Metastable Conditions in Cavitating Flows

Pietro Marco Congedo*, Maria Giovanna Rodio*, Gianluca Geraci†, Gianluca Iaccarino† and Remi Abgrall‡

200, Rue de la Vieille Tour, 33400 Talence, France
**INRIA Bordeaux-Sud-Ouest*

488 Escondido Mall, CA 94305-3035, USA
†Stanford University

Winterthurerstrasse 190, CH-8057 Zürich, Switzerland
‡Universität Zürich

The main objective of this work is to assess the influence of some relevant uncertainties on the variability of the cavitation structures in metastable conditions. First, the impact of physical model uncertainties is addressed, in particular focused on the thermodynamic model, since the proposed transition modelling is parameter-free. Secondly, physical modelling and experimental uncertainties are characterized in some well-known numerical experiments for two-phase flows with cavitation. Then, these uncertainties are propagated through a two-phase numerical solver for evaluating the impact on the predictive character of the numerical solution. In particular, the variability of some quantities of interest, such as the cavitation length, mixture pressure, is computed, thus permitting to evaluate the robustness of the physical model.

1. Introduction

The fast acceleration of a liquid produces several phenomena, as for instance, a rarefaction wave followed by the creation of vapour/gas bubbles. This is called Cavitation, and is one of the most inauspicious events in a mechanical device, since it can generate destructive and irreversible effects. In fact, the consequences of cavitation (see Chapter 3 of Ref. 1), such as noise, performance device reduction and wall corrosion, are extremely negative.

For these reasons, this phenomenon requires a good theoretical comprehension and, as consequence, an accurate predictive physical model is required. In the presence of a strong rarefaction wave in a superheated liquid (the temperature is higher than the saturated one at the final pressure of the expanded state), metastable states could be reached.[2,3] This means that, in these conditions, the liquid phase releases its metastable energy, producing vapour and, thus, evolving in a new equilibrium state. These events occur often in several applications, such as liquid flows around hypervelocity projectiles or in fuel injector systems, yielding dramatic consequences in terms of efficiency and maintenance.[2] Several numerical models have been proposed to simulate an interface problem. Principally two classes can be identified: (1) methods considering interfaces as true discontinuities and (2) methods allowing a numerical diffusion at the interface.

Methods belonging to the second class treat the interface like an artificial transition region where the thermodynamic conditions are unknown and they are known as *Interface Diffusive Models.* A hierarchy of models exists, with the numbers of equations ranging from seven to three only. The full non-equilibrium seven-equation models are the most complete. For both fluids, it contains equations for the mass, momentum and energy, and the seventh equation describes the topology of the flow. These models can take into account the physical details occurring in the cavitation phenomenon such as mass exchange, thermal transfer and surface tension. Anyway, these models are numerically complex, and more recent works have proposed the introduction of relaxation procedures in order to reduce the system of equation.[4–6] The most important reference about the treatment of heat and mass transfer terms, for the Interface Diffusive models, is the work of Saurel *et al.*[2] The authors proposed a phase transition modeling for a single velocity and single pressure model (five equations model) relying on a thermo-chemical relaxation. In fact, by obtaining the equality of temperature and chemical potential for the two phases, this relaxation assures the evolution of liquid phase from a metastable state to an equilibrium state on the saturation curve. In recent years, a few works have been proposed for treating phase transition modeling in Interface Diffusive models.[2,7–12] Following the approach of Ref. 2, Zein *et al.*[8] proposed a heat and mass transfer model for the six equation system (single velocity model), based on three different relaxations of pressure, temperature and chemical potential, respectively. Recently, the same approach has been used by Wang *et al.*[11] and Daude *et al.*[10] Nevertheless, the original relaxation procedure of Ref. 2 and the successive modifications are affected by some numerical issues.

In order to preserve the positivity of volume fraction and density, it might be necessary to integrate the system using time steps that are fractions of the hydrodynamic time step, thus yielding an increase of the computational cost (see Ref. 8). Recently, Pelanti and Shyue[9] presented a new numerical treatment of heat and mass transfer terms for a six equation two-phase model. As in Ref. 13, they proposed, in an original way, a new procedure in order to guarantee the positivity of the system, based on the total entropy maximization, when the solution coming from the thermo-chemical relaxation step, is not physically admissible.

In this paper, the approach proposed in Ref. 14 has been followed: the treatment of heat and mass transfer terms proposed in Ref. 2 for a five equation model is coupled with the procedure for solution admissibility of Ref. 9, thus preserving the positivity of the solution and reducing consistently the computational cost. As a consequence, this method combines a good accuracy and a reduced computational cost, as demonstrated in the following by performing several comparisons with experimental data and other numerical results well-known in the literature.

An accurate description of the cavitation phenomenon requires the modelling of the heat and mass transfer between the two-phases, the thermodynamic characterization of the phase change and the formulation of a complete model permitting to consider compressibility effects. For all these reasons, numerical simulation is very challenging due to the complexity of formulating a robust numerical scheme. An open question is still related to the assessment of the cavitation model. In particular, numerical solution and some quantities of interest seem extremely sensitive to some specific parameters used for defining properly the transition modelling. Despite the use of stochastic methods applied to the numerical simulation in fluid mechanics being more and more diffused, only few studies exist concerning the application of uncertainty quantification tools to cavitating flows. Li *et al.*[15] proposed a Markov stochastic model to reproduce the random behaviour of cavitation bubble(s) near compliant walls. Fariborza *et al.*[16] proposed an empirical model for the time-discrete stochastic nucleation of intergranular creep cavities. They assumed nucleation to occur randomly in time, with the temporal behaviour being governed by an inhomogeneous Poisson process. Giannadakis *et al.*[17] described the bubble breakup in Lagrangian models using a stochastic Monte-Carlo approximation. This study was oriented on the particular topic of cavitation in the Diesel nozzle holes. Mishra *et al.*[18] introduced a model of cavitation coupled to deterministic and stochastic chemical reactions of solute chemical species. Wilczynski[19] and Goel *et al.*[20] performed an uncertainties-based study on some hydrodynamic cavitation model parameters. In particular, Wilczynski[19] applied a stochastic model to capture the interaction of turbulent pressure field on cavitation nuclei population. Moreover, Goel *et al.*[20] performed a sensitivity analysis on several empirical parameters used typically in two-phase models. This study was performed using a finite differences method. In this case, input data uncertainty characterization is not required for the sensitivity analysis, that can be performed basing only on the mathematical form of the model. A recent work[21] presents a sensitivity study of the cavitation and turbulence models, where different combinations of empirical coefficients are considered. Recently, Rodio & Congedo[22] have proposed a study about the impact of various sources of uncertainty (on the cavitation model and on the inlet conditions) on the prediction of steady cavitating flows by coupling a non-intrusive Polynomial Chaos stochastic method with a cavitating CFD solver.

The main objective of this work is to assess the influence of some uncertainties on the variability of the cavitation structures in metastable conditions. For this reason, the impact of thermodynamic model uncertainties and of the experimental uncertainties are considered during the analysis of some well-known numerical experiments for two-phase flows with cavitation. Then, these uncertainties are propagated through a two-phase numerical solver for evaluating the impact on the predictive character of the numerical solution. In particular, the variability of some quantities of interest, such as the cavitation length, mixture pressure, is computed, thus permitting to evaluate the robustness of the physical model.

The paper is organized as follows. Section 2 presents the five-equation model with heat and mass transfer. Section 3 illustrates the main ingredients of the proposed approach for simulating cavitating flows. In Sec. 4, the thermodynamic closure is illustrated for each phase. In Sec. 5, some details concerning the Uncertainty Quantification method are provided. Finally, Sec. 6 illustrates several numerical results obtained on the most used numerical configuration in the literature.

2. A Synthetic Sketch of the Model

We introduce here the general formulation of a five-equation model, with mechanical relaxation (single pressure and single velocity for both phases) with heat and mass transfer term:[2]

$$\begin{aligned}
&\frac{\partial \alpha_1}{\partial t} + v\frac{\partial \alpha_1}{\partial x} = K\frac{\partial v}{\partial x} + \eta Q + \frac{\rho}{\rho_I}\dot{Y},\\
&\frac{\partial \alpha_1\rho_1}{\partial t} + \frac{\partial(\alpha_1\rho_1 v)}{\partial x} = \rho\dot{Y},\\
&\frac{\partial \alpha_2\rho_2}{\partial t} + \frac{\partial(\alpha_2\rho_2 v)}{\partial x} = -\rho\dot{Y},\\
&\frac{\partial \rho v}{\partial t} + \frac{\partial(\rho v^2 + P)}{\partial x} = 0,\\
&\frac{\partial \rho E}{\partial t} + \frac{\partial(\rho E + P)v}{\partial x} = 0,
\end{aligned} \tag{4.1}$$

where α_1, ρ_k, v, P, e_k, Q and $\dot{Y}$ are the gas volume fraction, the phase density, the mixture velocity, the mixture pressure, the phase internal energy , the heat term and the mass term, respectively. $E = e + \frac{1}{2}v^2$ is the mixture total energy and $e = (\alpha_1\rho_1 e_1 + \alpha_2\rho_2 e_2)/\rho$ is the mixture internal energy. Note that $K = \frac{\alpha_1\alpha_2(\rho_2 c_2^2 - \rho_1 c_1^2)}{(\alpha_1\rho_2 c_2^2 + \alpha_2\rho_1 c_1^2)}$. The unknowns are then: α_1, ρ_k, v, P, e_k, Q and $\dot{Y}$. This means that, for closing the system, two equations of state are needed (energy as a function of pressure and density for both phases) and a cavitation model is used to define Q and $\dot{Y}$. This is detailed in Sec. 3.

3. Cavitation Model and Global Numerical Method

In order to find the solution of system 4.1, a splitting procedure is adopted by performing the following three steps:

- *Step 1*: compute the numerical solution of the hyperbolic part of system (4.1) without heat and mass transfer term source. The discrete equation method (DEM)[23] is applied, which allows to treat the approximation of a non-conservative term, when the velocity $\vec{v}$ and K are simultaneously discontinuous.[5,23,24]
- *Step 2*: update the solution of system (4.1), by solving the temporal ODEs system with the heat and mass transfer terms. This consists in applying a thermo-chemical relaxation that allows, from a metastable state, to attain a new equilibrium solution.
- *Step 3*: the thermo-chemical relaxation does not guarantee that the positivity of the solution be preserved, so a positivity check of the solution is performed.

By adopting the splitting method, it is assumed that the characteristic time of a mechanical relaxation, $1/\mu$, is much smaller than the characteristic time scales $1/\theta$ and $1/\nu$ of heat and mass transfer terms, respectively. This means that thermal and chemical relaxation occur in a pressure equilibrium condition. In the following sections, the three steps are described in detail.

3.1. *Step 1: Numerical solution of a five-equation model without heat and mass transfer*

The numerical scheme for the solution of the five-equation model has been proposed by Abgrall and Saurel[23] and, successively, modified in Abgral and Perrier.[24] The original scheme has not been modified, so, in the following section, we present a summary description. We assume that at time t, the computational domain Ω is divided into the cells $C_i =]x_{i-1/2}, x_{i+1/2}[$. At a time $t = t + s$ (with s small), the interface in $x_{i+1/2}$ moves at a velocity $\sigma_{i+1/2}$ and the interface in $x_{i-1/2}$ moves at a velocity $\sigma_{i-1/2}$. As a consequence, the cell C_i evolves in $\bar{C}_i =]x_{i-1/2} + s\sigma_{i-1/2}, x_{i+1/2} + s\sigma_{i+1/2}[$ (see Fig. 4.1). The cell may be either smaller or larger than the original ones C_i, depending on the signs of the velocities. So, the Godunov scheme is no longer applied on the mesh cells, but on the modified and non-uniform cells constructed according to the position of the interface (see Fig. 4.1).[23] Then, we introduce the characteristic function X^k of the phase Σ_k. The function $X^k(x,t)$ is equal to 1, if x lies in the fluid Σ_k at time t and 0 otherwise.

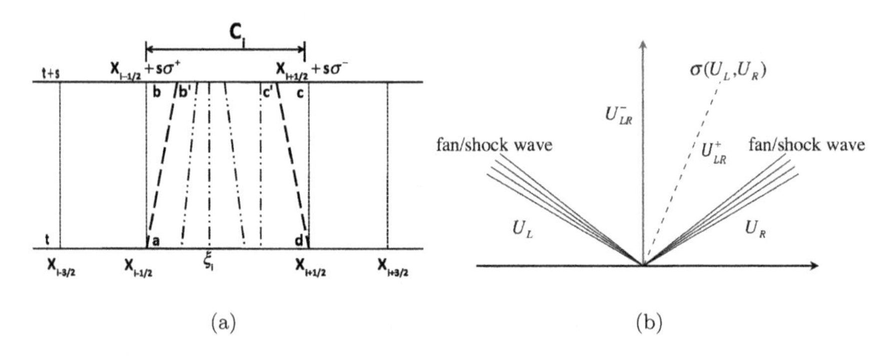

Fig. 4.1. (a) Subdivision of the computational domain. (b) The various states in the Riemann problem between states U_L and U_R.

Then, we denote with $F(U_L, U_R)$ the Godunov numerical flux between the states U_L and U_R, and with $F^{lag}(U_L, U_R)$ the flux across the contact interface between the states U_L and U_R (see Fig. 4.1). The relation between the two fluxes is equal to:

$$F^{lag}(U_L, U_R) = F(U_{LR}^+) - \sigma(U_L, U_R)U_{LR}^+ = F(U_{LR}^-) - \sigma(U_L, U_R)U_{LR}^-, \tag{4.2}$$

where the superscripts $\pm$ denote the state on the right and on the left of the contact discontinuity, as in Fig. 4.1.

In Ref. 23, a semi-discrete equation for the discretization of a seven-equation model has been developed. Then, in Ref. 24, an asymptotic development has been applied on the numerical approximation of Ref. 23, obtaining the semi-discrete equation for the reduced model (see Eq. (4.1)). This last one is used in this work and its formulation is the following:

$$\begin{aligned}
\frac{\partial \alpha_1}{\partial t} &= FT(U_1) + \frac{\alpha_1\alpha_2}{\alpha_2\rho_1 a_1^2 + \alpha_1\rho_2 a_2^2}\left\{\frac{FT(U_8)}{\alpha_2\rho_2\chi_2} - \frac{u_2 FT(U_7)}{\alpha_2\rho_2\chi_2}\right. \\
&\quad + \frac{\frac{u_2{}^2}{2} - e_2 - \rho_2\kappa_2}{\alpha_2\rho_2\chi_2}FT(U_6) + \frac{\rho_2^2\kappa_2 FT(U_1)}{\alpha_2\rho_2\chi_2} - \frac{FT(U_4)}{\alpha_1\rho_1\chi_1} \\
&\quad \left. + \frac{u_1 FT(U_3)}{\alpha_1\rho_1\chi_1} - \frac{\frac{u_1{}^2}{2} - e_1 - \rho_1\kappa_1}{\alpha_1\rho_1\chi_1}FT(U_2) - \frac{\rho_1^2\kappa_1 FT(U_5)}{\alpha_1\rho_1\chi_1}\right\}, \\
\frac{\partial \alpha_1\rho_1}{\partial t} &= FT(U_2), \\
\frac{\partial \alpha_2\rho_2}{\partial t} &= FT(U_6), \\
\frac{\partial \rho u}{\partial t} &= FT(U_3) + FT(U_7), \\
\frac{\partial \rho E}{\partial t} &= FT(U_4) + FT(U_8).
\end{aligned} \tag{4.3}$$

The vector $FT(U_j)$, with $j = 1, \ldots, 8$, is defined by

$$\begin{aligned}
FT(U_j) &= \frac{1}{\Delta x}\,\mathcal{E}\big(X(x_{i+1/2}, t)F(U_{i+1/2}^\star) - X(x_{i-1/2}, t)F(U_{i-1/2}^\star)\big) \\
&\quad + \frac{1}{\Delta x}\left(\mathcal{E}([X]_{j=0})F^{lag}(U_i^-, U_{i-1}^+) - \mathcal{E}([X]_{j=N})F^{lag}(U_i^+, U_{i+1}^-)\right),
\end{aligned} \tag{4.4}$$

where $U_{i+1/2}^\star$ (or $U_{i-1/2}^\star$) denotes the solution of Riemann problem between U_i^+ and U_{i+1}^- (respectively, U_{i-1}^+ and U_i^-). The quantities $[X]_{j=0}$ and $[X]_{j=N}$ are the jumps of X at the beginning and at the end of the computational cell, respectively.

The correspondence of the semi-discrete system (4.3) with the model (4.1) has been demonstrated in Ref. 24. This method assumes initially two different thermodynamic states of phases, and at the end of the time step, the states have relaxed to a mechanical equilibrium. On the contrary, a direct discretization of the system (4.1) assumes directly that the initial pressure and velocity of the phases are equal. The advantages of the DEM method are summarized in Refs. 23 and 24. Then, conservative and non-conservative terms of the system (4.3) can be developed assuming the four possible types of fluid phase discontinuities (Table 4.1).

Table 4.1. The various flow configurations at cell boundary $i+1/2$.

Flow patterns	Jump indicator	Flux indicator
$\Sigma_1 - \Sigma_2$	$[X]_{1,1} = 0$	$\left(\beta^{(1,2)}_{i+1/2}\right) = \begin{cases} 1 & \text{if } \sigma(U^l_i, U^r_{i+1}) \geq 0 \\ -1 & \text{if } \sigma(U^l_i, U^r_{i+1}) < 0 \end{cases}$
$\Sigma_1 - \Sigma_1$	$[X]_{1,2} = \begin{cases} -1 & \text{if } \sigma(1,2) > 0 \\ 0 & \text{otherwise} \end{cases}$	1
$\Sigma_2 - \Sigma_1$	$[X]_{2,1} = \begin{cases} 1 & \text{if } \sigma(2,1) > 0 \\ 0 & \text{otherwise} \end{cases}$	$\left(\beta^{(2,1)}_{i+1/2}\right) = \begin{cases} 1 & \text{if } \sigma(U^l_i, U^r_{i+1}) \geq 0 \\ -1 & \text{if } \sigma(U^l_i, U^r_{i+1}) < 0 \end{cases}$
$\Sigma_2 - \Sigma_2$	$[X]_{2,2} = 0$	0

The terms of the vector $FT(U_j)$ of Eq. (4.3) can be computed as follows:

$$\begin{aligned}\mathcal{E}\big(X(x_{i+\frac{1}{2}}, t)F(U^{\star}_{i+\frac{1}{2}})\big) &= \mathcal{P}_{i+\frac{1}{2}}(\Sigma_1, \Sigma_1)F(U^{(1)}_i, U^{(1)}_{i+1}) \\ &\quad + \mathcal{P}_{i+\frac{1}{2}}(\Sigma_1, \Sigma_2)\big(\beta^{(1,2)}_{i+\frac{1}{2}}\big)F(U^{(1)}_i, U^{(2)}_{i+1}) \\ &\quad + \mathcal{P}_{i+\frac{1}{2}}(\Sigma_2, \Sigma_1)\big(\beta^{(2,1)}_{i+\frac{1}{2}}\big)F(U^{(2)}_i, U^{(1)}_{i+1}),\end{aligned}$$

$$\begin{aligned}\mathcal{E}\big(X(x_{i-\frac{1}{2}}, t)F(U^{\star}_{i-\frac{1}{2}})\big) &= \mathcal{P}_{i-\frac{1}{2}}(\Sigma_1, \Sigma_1)F(U^{(1)}_{i-1}, U^{(1)}_{i}) \\ &\quad + \mathcal{P}_{i-\frac{1}{2}}(\Sigma_1, \Sigma_2)\big(\beta^{(1,2)}_{i-\frac{1}{2}}\big)F(U^{(1)}_{i-1}, U^{(2)}_{i}) \\ &\quad + \mathcal{P}_{i-\frac{1}{2}}(\Sigma_2, \Sigma_1)\big(\beta^{(2,1)}_{i-\frac{1}{2}}\big)F(U^{(2)}_{i-1}, U^{(1)}_{i}),\end{aligned}$$

$$\begin{aligned}\mathcal{E}\big([X]_N F^{lag}(U^{N(w)}_i, U^{-}_{i+1})\big) &= \mathcal{P}_{1+1/2}(\Sigma_1, \Sigma_2)\big(\beta^{(1,2)}_{i+1/2}\big)F^{lag}(U^{(1)}_i, U^{(2)}_{i+1}) \\ &\quad - \mathcal{P}_{1+1/2}(\Sigma_2, \Sigma_1)\big(\beta^{(2,1)}_{i+1/2}\big)F^{lag}(U^{(2)}_i, U^{(1)}_{i+1}),\end{aligned}$$

$$\begin{aligned}\mathcal{E}\left([X]_0 F^{lag}(U^{+}_{i-1}, U^{0}_i)\right) &= -\mathcal{P}_{1-1/2}(\Sigma_1, \Sigma_2)\big(\beta^{(1,2)}_{i-1/2}\big)F^{lag}(U^{(1)}_{i-1}, U^{(2)}_{i}) \\ &\quad + \mathcal{P}_{1-1/2}(\Sigma_2, \Sigma_1)\big(\beta^{(2,1)}_{i+1/2}\big)F^{lag}(U^{(2)}_{i-1}, U^{(1)}_{i}),\end{aligned}$$

where χ_k and κ_k are defined as follows:

$$\chi_k = \left(\frac{\partial e_k}{\partial P_k}\right)_{\rho_k}, \quad \kappa_k = \left(\frac{\partial e_k}{\partial \rho_k}\right)_{P_k}, \tag{4.5}$$

e_k is the phase internal energy.

It remains to evaluate the probabilities $\mathcal{P}_{i\pm1/2}(\Sigma_p, \Sigma_q)$.[23] For simplicity, we only show the final formulation for $i+1/2$:

$$\mathcal{P}_{i+1/2}(\Sigma_1, \Sigma_1) = \min\left(\alpha_i^{(1)}, \alpha_{i+1}^{(1)}\right),$$
$$\mathcal{P}_{i+1/2}(\Sigma_2, \Sigma_1) = \max\left(\alpha_i^{(2)} - \alpha_{i+1}^{(2)}, 0\right),$$
$$\mathcal{P}_{i+1/2}(\Sigma_1, \Sigma_2) = \min\left(\alpha_i^{(2)}, \alpha_{i+1}^{(2)}\right),$$
$$\mathcal{P}_{i+1/2}(\Sigma_1, \Sigma_2) = \max\left(\alpha_i^{(1)} - \alpha_{i+1}^{(1)}, 0\right),$$

where Σ_k indicates the phase, with $k = 1, 2$. The numerical flux $F(U)$ is approximated via an approximate Riemann solver which can define the contact speed $\sigma(U_L, U_R)$, allowing to define also the Lagrangian flux F^{lag} (see Eq. (4.2)). In this paper, we have used the relaxation solver for all computations.[24] A second order has been applied by means of a predictor-corrector approach that is an extension to a multiphase flows of the MUSCL method shown in Ref. 23. It has been extensively explained in Ref. 23. The time step is constrained by $|\lambda_{max}| \frac{\Delta x}{\Delta t} \leqslant \frac{1}{2}$.

The semi-discrete equation (4.3) is solved in order to find the solution of the hyperbolic system with mechanical equilibrium. This strategy allows to start from two completely different thermodynamic states as in the seven-equation models, but finally, to find the mechanical equilibrium solution of a five equation model.

We denote with superscript "0" the initial quantities at time t=0 and with superscript $\star$ the quantities computed by solving Step 1. In order to close the system (4.1), an equation of state (EOS) for each phase and for the mixture is used.[25]

3.2. *Step 2: Numerical solution of the temporal ODEs with heat and mass transfer terms*

After considering the hydrodynamic evolution, let us focus now on the numerical treatment of heat and mass transfer terms. So, we consider the system (4.1) without the convective terms. The following ODEs equation system allows to let evolve the solution, considering the new source terms:

$$
\begin{aligned}
\frac{\partial \alpha_1}{\partial t} &= \eta Q + \frac{\rho}{\rho_I}\dot{Y} := S_{\alpha_1}, \\
\frac{\partial \alpha_1 \rho_1}{\partial t} &= \rho \dot{Y} := S_{Y_1}, \\
\frac{\partial \alpha_2 \rho_2}{\partial t} &= -\rho \dot{Y} := -S_{Y_1}, \\
\frac{\partial \rho v}{\partial t} &= 0, \\
\frac{\partial \rho E}{\partial t} &= 0.
\end{aligned}
\tag{4.6}
$$

where: $Q = \theta(T_2 - T_1)$, $\dot{Y} = \nu(g_2 - g_1)$ and $\eta = \frac{\alpha_1\alpha_2}{\alpha_2\rho_1 c_1^2 + \alpha_1 \rho_2 c_2^2}(\frac{\Gamma_1}{\alpha_1} + \frac{\Gamma_2}{\alpha_2})$. The Gibbs free energy (chemical potential), g_k, are defined for a stiffened gas equation of state, as follows (see Ref. 26 for more details):

$$
g_k(P, T_k) = (\gamma_k C v_k - q')T_k - C v_k T_k \log\left(\frac{T_k^{(\gamma_k)}}{(P + P_{\infty,k})^{\gamma_k - 1}}\right).
$$

The interface density has been defined in Ref. 2 and it has been determined assuming an isentropic behavior of acoustic waves, as follows:

$$
\rho_I = \frac{\frac{\rho_1 c_1^2}{\alpha_1} + \frac{\rho_2 c_2^2}{\alpha_2}}{\frac{c_1^2}{\alpha_1} + \frac{c_2^2}{\alpha_2}}. \tag{4.7}
$$

Parameters θ and ν are the thermal and chemical relaxation parameters, respectively.

To identify the liquid/vapour interface, the method proposed in Ref. 2 and then in Refs. 8 and 9 is used. In particular, the relaxation parameters θ and ν are set to zero far from the interfaces, and they are taken as infinite in order to fulfil equilibrium interface conditions with mass transfer:

$$
\theta, \nu = \begin{cases} +\infty & \text{if } \epsilon \leq \alpha_1 \leq 1 - \epsilon, \\ 0 & \text{otherwise,} \end{cases} \tag{4.8}
$$

where ϵ is a very low value. In order to define this value, we first remember that a limit of this model is that a phase can not completely disappear. This means that in all cells, the two phases always coexist. Anyway, a cell is considered filled by a pure fluid, when its volume and mass fraction is equal to $1 - \epsilon_1$, with typically $\epsilon_1 = 10^{-8}$. In other words, in all cells, the gas phase can have a volume fraction value included in $\epsilon_1 < \alpha_g < 1 - \epsilon_1$ and the liquid phase a complementary volume fraction, $\alpha_l = 1 - \alpha_g$, in order to assure that also the other phase exists.

As a consequence, for identifying the interface, $\epsilon = 10^{-6}$ is taken, and the interface corresponds to mixture cells when volume fractions range is between ϵ and $1 - \epsilon$. If ϵ and ϵ_1 are too close, evaporation may occur too early and not only in the interfacial zone.

Moreover, we impose that the mass transfer is allowed only if metastable state is fulfilled, i.e. if $T_k > T_{sat}(P^\star)$. The saturation temperature has been defined in Sec. 4.

Let us denote with superscript '$\star\star$' the final quantities computed by solving Step 1. In Table 4.2, the input and output quantities of the second step and the variables that are not influenced by the heat or mass transfer are reported. In fact, from system (4.8), it is clear that thanks to mass conservation, the mixture density is constant ($\rho^{\star\star} = \rho^\star$) and thus, the velocity ($v^{\star\star} = v^\star$) and the mixture total and internal energy ($E^{\star\star} = E^\star$ and $e^{\star\star} = e^\star$).

Remembering that $\alpha_2 = 1-\alpha_1$, the unknowns of the system (4.6) are α_1, ρ_k, v, P, e_k, Q and $\dot{Y}$. Then, in order to close the system, two ingredients are required: (i) an equation of state for defining the internal energy as function of pressure and density for both phases (see Sec. 4) (ii) a thermo-chemical relaxation for finding Q and $\dot{Y}$. The thermo-chemical relaxation is described in the following sections (Sec. 3.2.1).

Table 4.2. Variables used in Step 2. $\rho = (\alpha_1\rho_1) + (\alpha_2\rho_2)$; $\rho e = (\alpha_1\rho_1 e_1) + (\alpha_2\rho_2 e_2)$; $E = e + \frac{1}{2}v^2$.

Step 2 (Only at the interface: if $\epsilon_1 < \alpha_1 < 1 - \epsilon_1$ and $T_k > T_{sat}(P^\star)$)		
IN variables	OUT variables	Constant MIXTURE variables
$\alpha_1^\star$ ($\alpha_2^\star = 1 - \alpha_1^\star$)	$\alpha_1^{\star\star}$ ($\alpha_2^{\star\star} = 1 - \alpha_1^{\star\star}$)	
$\rho_1^\star$ and $\rho_2^\star$	$\rho_1^{\star\star}$ and $\rho_2^{\star\star}$	
$v^\star$	$v^{\star\star}$	$\rho^{\star\star} = \rho^\star$
$P^\star$	$P^{\star\star}$	$e^{\star\star} = e^\star$
$e_1^\star$ and $e_2^\star$	$e_1^{\star\star}$ and $e_2^{\star\star}$	$E^{\star\star} = E^\star$
$T_1^\star$ and $T_2^\star$	$T^{\star\star}$	
$g_1^\star$ and $g_2^\star$	$g^{\star\star}$	

3.2.1. *Stiff thermo-chemical solver for computing Q and $\dot{Y}$*

It is well-known that, physically, the single phase in metastable state evolves in a new equilibrium liquid/vapour state with a heat and mass exchange. Thus, the mass transfer can stop, only when a new state on the saturation curve is attained, since the liquid and its vapour are in equilibrium. In order to estimate if the two phases are in equilibrium on the saturation curve, it should be assured that, in all time steps, the system achieves the temperature and Gibbs free energy equilibrium between the phases ($T_1 = T_2$ and $g_1 = g_2$).

For solving this issue, a procedure similar to Ref. 2 is used. It can be demonstrated that the temperature and the Gibbs time derivatives can be written in terms of Q and $\dot{Y}$, using the mass equations and the energy mixture equation of system (4.6) as follows:

$$\begin{cases} \dfrac{\partial \Delta T}{\partial t} = AQ + B\dot{Y}, \\ \dfrac{\partial \Delta g}{\partial t} = A'Q + B'\dot{Y}, \end{cases} \tag{4.9}$$

where A, B, A' and B' should be defined. The simplest numerical approximation of Eq. (4.9) can be written as follows:

$$\begin{aligned} \frac{\partial \Delta T}{\partial t} &= \frac{\Delta T^{n+1} - \Delta T^n}{\Delta t} = \frac{0 - \Delta T^n}{\Delta t} = A^n Q^n + B^n \dot{Y}^n, \\ \frac{\partial \Delta g}{\partial t} &= \frac{\Delta g^{n+1} - \Delta g^n}{\Delta t} = \frac{0 - \Delta g^n}{\Delta t} = A'^n Q^n + B'^n \dot{Y}^n. \end{aligned} \tag{4.10}$$

Assuming that thermodynamic equilibrium is reached at the end of each time step, the temperature and Gibbs differences, ΔT^{n+1} and Δg^{n+1}, in Eq. (4.10) can be imposed equal to zero. Thus, finally, we can determine the heat and mass transfer terms that are computed as follows:

$$\begin{cases} Q = -\dfrac{B'}{AB' - A'B}\dfrac{(\Delta T)^n}{\Delta t} + \dfrac{B}{AB' - A'B}\dfrac{(\Delta g)^n}{\Delta t}, \\ \dot{Y} = \dfrac{A'}{AB' - A'B}\dfrac{(\Delta T)^n}{\Delta t} + \dfrac{A}{AB' - A'B}\dfrac{(\Delta g)^n}{\Delta t}. \end{cases} \tag{4.11}$$

To compute Q and $\dot{Y}$, A, B, A' and B' for a five-equation model should be computed. After some manipulations of temporal derivatives of temperature and Gibbs free energy, by exploiting the mass conservation equations of each phase and the mixture total energy conservation equation in the system (4.6) (see Ref. 5 for more details), we can define the coefficients A,

B, A', B' as follows:

$$
\begin{aligned}
A &= -\xi\left[\left(\frac{T_2 D_2 Y_1}{T_1 D_1 Y_2} - 1\right)\left(Z_1\eta - \frac{T_2 C_{v_2}}{\rho} C_1\eta\right) + \frac{T_2}{\alpha_2\rho_2} C_1\eta\right], \\
B &= -\xi\left(\frac{T_2 D_2 Y_1}{T_1 D_1 Y_2} - 1\right)\left[\theta_1 + \frac{T_2 C v_2}{\rho}\left(C_2 - \frac{\rho}{\rho_I}\right)\right] + \frac{T_2}{\alpha_2\rho_2}\left(C_2 - C_1\frac{\rho}{\rho_I}\right), \\
A' &= -\left(\frac{1}{\rho_2} - \frac{1}{\rho_1}\right)\frac{c_1^2}{\alpha_1\gamma_1}\rho_1\eta + \frac{S_2 T_2}{\alpha_2\rho_2} C_1\eta - \xi\delta_1\left(Z_1\eta - \frac{T_2 C_{v_2}}{\rho} C_1\eta\right), \\
B' &= \left(\frac{1}{\rho_2} - \frac{1}{\rho_1}\right)\frac{c_1^2}{\alpha_1\gamma_1}\left(\rho - \rho_1\frac{\rho}{\rho_I}\right) - \frac{S_2 T_2}{\alpha_2\rho_2}\left(C_2 - C_1\frac{\rho}{\rho_I}\right) \\
&\quad - \xi\delta_1\left[\theta_1 + \frac{T_2 C_{v_2}}{\rho}\left(C_2 - \frac{\rho}{\rho_I}\right)\right],
\end{aligned}
\tag{4.12}
$$

where $\xi = \frac{T_1 D1}{Y_1(C_{v_1} T_1 D_1 + C_{v_2} T_2 D_2)}$.

3.3. *Step 3: Mass fraction and density positivity*

The approximation of heat and mass transfer terms allows the calculation of source terms of volume fraction and mass conservative equations, but there is no guarantee that positivity of the solution be preserved. Assuming an evaporation process for example, the mass source terms estimated can be larger than the liquid mass fraction that can evaporate. So, as in Ref. 2, a limitation is placed on the source terms, by determining the maximum admissible values, as follows:

$$
S_{max,\alpha_1} = \begin{cases} \dfrac{1-\alpha_1}{\Delta t} & \text{if } S_{max} > 0, \\ \dfrac{-\alpha_1}{\Delta t} & \text{otherwise,} \end{cases} \qquad S_{max,Y_1} = \begin{cases} \dfrac{(1-\alpha_1)\rho_2}{\Delta t} & \text{if } S_{max} > 0 \\ \dfrac{-\alpha_1\rho_1}{\Delta t} & \text{otherwise.} \end{cases}
$$

Thus, if $|S_{max,\alpha_1}| > |S_{\alpha_1}|$ and $|S_{max,Y_1}| > |S_{Y_1}|$, the source terms is used.

Otherwise, if the limit value is not respected, a new method is introduced in order to avoid an integration of the system over a fractional hydrodynamic time step, as it is done in Refs. 2 and 8. We assumed that the mixture is composed nearly of the species k that has the highest entropy, using an idea similar to Ref. 9. In particular, let us indicate with $\bigstar$ the variable value at the end of positivity system check control:

- if $s_g^\bigstar > s_l^\bigstar$, then the vapour volume fraction, $\alpha^{\star\star}$, is fixed to $1-\epsilon_1$,
- otherwise, if $s_g^\bigstar < s_l^\bigstar$, then $\alpha^{\star\star}$ is fixed to ϵ_1,

where, as explained in Sec. 3.2, $\epsilon_1 = 10^{-8}$.

Knowing $\alpha^{\star\star}$, it can be used in the following system in order to find the new equilibrium pressure, $P^{\star\star}$, and temperature, $T^{\star\star}$, and, then the other thermodynamic variables (a stiffened equation of state is assumed):

$$\begin{aligned}
\rho^{\star\star} &= (\alpha_1\rho_1)^{\star\star} + (\alpha_2\rho_2)^{\star\star} = \rho^{\star}, \\
(\rho e)^{\star\star} &= (\alpha_1\rho_1 e_1)^{\star\star} + (\alpha_2\rho_2 e_2)^{\star\star} = (\rho e)^{\star}, \\
\rho_k^{\star\star}(P^{\star\star}, T^{\star\star}) &= \frac{P^{\star\star} + P_{\infty,k}}{Cv_k\Gamma_k T^{\star\star}}, \\
(\rho e)_k^{\star\star}(P^{\star\star}, T^{\star\star}) &= \frac{P^{\star\star} + P_{\infty,k}\gamma_k}{\Gamma_k} + \rho_k^{\star\star} q,
\end{aligned} \tag{4.13}$$

where $\Gamma = \gamma - 1$. By replacing the phase density, $\rho_k^{\star\star}$, and the phase energy $(\rho e)_k^{\star\star}$, in the first two equations of system (4.13), we obtain a single quadratic equation for $P^{\star\star}$:

$$\begin{aligned}
T^{\star\star} &= \frac{\alpha_1^{\star\star} Cv_2\Gamma_2(P^{\star\star} + P_{\infty,1}) + \alpha_2^{\star\star} Cv_1\Gamma_1(P^{\star\star} + P_{\infty,2})}{\rho^{\star} Cv_1\Gamma_1 Cv_2\Gamma_2}, \\
0 &= (P^{\star\star})^2 + bP^{\star\star} + d,
\end{aligned} \tag{4.14}$$

and:

$$\begin{aligned}
b &= \frac{F_1}{F_2} + \frac{q_1(\alpha_1^{\star\star} C_{v_2}\Gamma_2) + q_2(\alpha_2^{\star\star} C_{v_1}\Gamma_1)}{Z_2} \\
&\quad + F_2 \frac{P_{\infty,1}(\alpha_1^{\star\star} C_{v_2}\Gamma_2) + P_{\infty,2}(\alpha_2^{\star\star} C_{v_1}\Gamma_1)}{\rho^{\star}\Gamma_1\Gamma_2}, \\
d &= \frac{F_1}{Z_2} \frac{P_{\infty,1}(\alpha_1^{\star\star} C_{v_2}\Gamma_2) + P_{\infty,2}(\alpha_2^{\star\star} C_{v_1}\Gamma_1)}{\rho^{\star}\Gamma_1\Gamma_2} \\
&\quad + \frac{P_{\infty,1} q_1(\alpha_1^{\star\star} C_{v_2}\Gamma_2) + P_{\infty,2} q_2(\alpha_2^{\star\star} C_{v_1}\Gamma_1)}{Z2},
\end{aligned} \tag{4.15}$$

where

$$F_1 = P_{\infty,1}\gamma_1 \frac{(\alpha_1^{\star\star} C_{v_2}\Gamma_2)}{C_{v_2}} + P_{\infty,2}\gamma_2 \frac{(\alpha_2^{\star\star} C_{v_1}\Gamma_1)}{C_{v_1}} - (\rho e)^{\star}\Gamma_1\Gamma_2,$$

$$F_2 = \alpha_1^{\star\star}\Gamma_2 + \alpha_2^{\star\star}\Gamma_1 \qquad \text{and} \qquad Z_2 = F_2 \frac{(\alpha_1^{\star\star} C_{v_2}\Gamma_2) + (\alpha_2^{\star\star} C_{v_1}\Gamma_1)}{\rho^{\star\star}\Gamma_1\Gamma_2}.$$

Then, by solving the single quadratic equation of $P^{\star\star}$, we select the physically admissible solution of the quadratic equation that maximizes the total entropy $s^{\star\star} = Y_1^{\star\star} s_1^{\star\star} + Y_2^{\star\star} s_2^{\star\star}$, where $Y_k^{\star\star} = (\alpha_k^{\star\star}\rho_k^{\star\star})/\rho^{\star\star}$.

This procedure allows to reduce the computational cost compared to Refs. 2 and 8: it is not necessary to integrate over a fractional hydrodynamic time step.

4. Thermodynamic Closure

As we have previously mentioned, we deal with pure fluid and artificial mixture zone, thus the EOS must be able to describe flows both in pure fluids and mixture zones.

In order to clarify this point, let us remember that, in the system 4.1, the last two equations are the momentum and energy conservative equation for the mixture, in which, a mixture pressure, P, appears. It is for this reason that we need a mixture EOS, in order to extrapolate the value of P.

In this section, the subscript k and m are used to indicate the phase variables and the mixture ones, respectively. The stiffened gas EOS is usually used for shock dynamics and its robustness for simulating two-phase flow with or without mass transfer has been amply demonstrated, see for example Refs. 23–25. It can be written as follows:

$$P_k(\rho_k, e_k) = (\gamma_k - 1)(e_k - q_k)\rho_k - \gamma_k P_{\infty,k}, \tag{4.16a}$$

$$e_k(\rho_k, T_k) = T_k c_{v,k} + \frac{P_{\infty,k}}{\rho_k} + q_k, \tag{4.16b}$$

$$s_k(P_k, T_k) = c_{v,k} ln \frac{T_k^{\gamma_k}}{(P_k + P_{\infty,k})^{(\gamma_k - 1)}} + q'_k, \tag{4.16c}$$

where P_k, ρ_k, e_k and s_k are the pressure, the density, the internal energy and the entropy of the phase, respectively. The polytropic coefficient γ_k is the constant ratio of specific heat capacities $\gamma_k = c_{p,k}/c_{v,k}$, $P_{\infty,k}$ is a constant reference pressure, q_k is the creation energy of the fluid at a given reference state and q' is a fluid specific constant. Moreover, T_k, $c_{v,k}$ and h_k are the temperature, the specific heat at constant volume and the enthalpy, respectively. For more details about the stiffened gas equation of state, we strongly suggest the reading of Refs. 2 and 26.

The speed of sound, defined as $c_k^2 = (\frac{\partial P_k}{\partial \rho_k})_s$ can be computed as follows:

$$c_k^2 = \gamma_k \frac{P_k + P_{\infty,k}}{\rho_k} = (\gamma_k - 1)c_{p,k}T_k, \tag{4.17}$$

where c_k^2 remains strictly positive (for $\gamma_k > 1$). It ensures the hyperbolicity of the system and the existence of a convex mathematical entropy.[27]

The EOS for the mixture can be easily obtained using the EOS of the single phase. The aim is now to obtain the mixture pressure. The starting point is the mixture energy equation, i.e. $\rho_m e_m = \alpha_1 \rho_1 e_1 + \alpha_2 \rho_2 e_2$, where 1 and 2 indicate the gas and the liquid phase. The energy of each phase,

$e_k = \frac{P_k + P_{\infty,k}\gamma_k}{\rho_k(\gamma_k - 1)}$, can be replaced to obtain the mixture total energy as a function of the phase pressures. Under pressure equilibrium, we obtain the following expression for the pressure mixture:

$$P_m(\rho_m, e_m, \alpha_k) = \frac{\rho_m(e_m - \frac{\alpha_1 \rho_1 q_1}{\rho_m} - \frac{\alpha_2 \rho_2 q_2}{\rho_m}) - \left(\frac{\alpha_1 \gamma_1 P_{\infty,1}}{\gamma_1 - 1} + \frac{\alpha_2 \gamma_2 P_{\infty,2}}{\gamma_2 - 1}\right)}{\frac{\alpha_1}{\gamma_1 - 1} + \frac{\alpha_2}{\gamma_2 - 1}}. \tag{4.18}$$

5. Forward Uncertainty Quantification Method

Concerning uncertainty propagation, a truncated Polynomial Chaos expansion (see Ref. 28) is computed. Using this non-intrusive uncertainty quantification tool means that a single deterministic computation is replaced with a whole set of such computations, each one of those being run for specific values of the uncertain conditions.

Polynomial Chaos (PC) expansions are derived from the original theory of Wiener on spectral representation of stochastic processes using Gaussian random variables. Let $\boldsymbol{\xi}$ be a vector of standard independent random variables $\xi_i, i = 1, 2, \ldots, n_\xi$. Any well-behaved process u (i.e. a second-order process, then with a finite variance) can be expanded in a convergent (in the mean square sense) series of the form

$$u(\mathbf{x}, t, \boldsymbol{\xi}) = \sum_{\alpha} u_\alpha(\mathbf{x}, t)\Psi_\alpha(\boldsymbol{\xi}), \tag{4.19}$$

where α are multi-indices, $\alpha = (\alpha_1, \alpha_2, \ldots, \alpha_n)$, with each component $\alpha_i = 0, 1, \ldots$, and Ψ_α are multivariate polynomial functions orthogonal with respect to the probability distribution function of the vector $\boldsymbol{\xi}$. Each Ψ_α is defined by a product of orthogonal polynomials $\Phi_i^{\alpha_i}(\xi_i)$, that is, $\Psi_\alpha(\boldsymbol{\xi}) = \prod_{i=1}^{n_\xi} \Phi_i^{\alpha_i}(\xi_i)$, where each $\Phi_i^{\alpha_i}$ is a polynomial of degree α_i, so that the degree of Ψ_α is $|\alpha|_1 = \sum_{i=1}^{n_\xi} \alpha_i$. A one-to-one correspondence exists between the choice of stochastic variable ξ_i and the polynomials $\Phi_i^{\alpha_i}(\xi_i)$. For instance, if ξ_i is a normal/uniform variable, the corresponding $\Phi_i^{\alpha_i}(\xi_i)$ are Hermite/Legendre polynomials of degree α_i. Coefficients $u_\alpha(x, t)$ are called the PC coefficients of the random process u and are obtained by

$$u_\alpha(\mathbf{x}, t) = \langle u(\mathbf{x}, t), \Psi_\alpha \rangle \left\| \Psi_\alpha \right\|^{-2}, \tag{4.20}$$

where the scalar product is defined by the expectation operator. For practical use, the PC expansions are truncated to degree No

$$u(\mathbf{x}, t, \boldsymbol{\xi}) = \sum_{|\alpha|_1 \leq \text{No}} u_\alpha(\mathbf{x}, t) \Psi_\alpha(\boldsymbol{\xi}). \tag{4.21}$$

The number of multivariate polynomials Ψ_α , that is, the dimension of the expansion basis, is related to the stochastic dimension n_ξ and the degree No of polynomials ; it is given by the formula $(n_\xi + \text{No})!/(n_\xi!\text{No}!)$.

Several approaches can be used to estimate PC coefficients. The approach used in this study is based on Gauss-Legendre quadrature nodes. From the PC expansion of the random process, it is then easy to derive its mean and variance and to estimate sensitivity information using the analysis of variance (ANOVA) decomposition.[29]

6. Results

In this section, we present some results about the uncertainty propagation of some physical and parameter uncertainties through a numerical code, which solves the model presented in the previous section.

Two well-known test-cases are selected, which are among the most used ones for the validation of compressible cavitation model : a two phase shock-tube, with the appearance of a metastable state and an evaporation wave, and a two-phase expansion tube test. The aim is to (i) show the robustness of cavitation and thermodynamic models; (ii) study the influence of model uncertainties in terms of prediction of some quantities of interest from a physical point of view.

Physical parameters are summarized in Table 4.3. A second order has been performed for all the numerical test-cases by means of a MUSCL scheme,[23] coupled with a Relaxation solver[24] and a Van Leer limiter. Note also that a mesh-convergence study has been done, even if it is not reported here for brevity.

Table 4.3. EOS coefficients for liquid and gas phases.

Test case	Fluid	SG EOS					
		γ	P_∞ $[Pa]$	C_p $[\frac{J}{KgK}]$	C_v $[\frac{J}{KgK}]$	q	q'
1	Liq. Dodecane	2.35	4×10^8	2534	1077	-755×10^3	0
Shock tube	Vap. Dodecane	1.025	0	2005	1956	-237×10^3	-24.4×10^3
2	Liq. Water	2.35	10^9	4267	1816	-1167×10^3	0
Expansion tube	Vap. Water	1.43	0	1487	1040	2030×10^3	-23.2×10^3

6.1. *Two-phase shock tube with mass transfer*

In this test-case, the shock tube is filled out with liquid dodecane on the left at high pressure $p_l = 10^8$ and vapour dodecane on the right at atmospheric pressure $p_g = 10^5$ (see Fig. 4.2). Note that, for numerical reasons, each chamber contains a weak volume of the other fluid ($\alpha_k = 10^{-8}$). The diaphragm is located at $x = 0.75$ m (the tube is 1 m long) and the results are shown at a time of $t = 473$ μs. The computation are performed by using a mesh of 1000 cells.

A comparison between the numerical solutions obtained with and without heat and mass transfer is performed for estimating the influence of these terms. Moreover, results are compared with the ones obtained in Ref. 9 (considering a pressure relaxation coupled with a pressure-thermo-chemical relaxation). These comparisons are reported in Figs. 4.3(a) and 4.3(b).

In particular, the influence of heat and mass transfer can be observed. In fact, an additional evaporation wave appears between the rarefaction wave (at $0.05 < x < 0.12$ m) and the contact discontinuity (at $0.87 < x < 0.9$ m) (see Figs. 4.3(a) and 4.3(b)). The evaporation wave determines a strong increase of velocity and an increase of pressure at $0.12 < x < 0.8$ m.

Fig. 4.2. Shock tube geometry.

An uncertainty quantification study is then performed by considering three uniform random variables, corresponding to an equation of state parameter, i.e. q', the left pressure and left temperature of the liquid phase at the initial state.

Because of the lack of knowledge concerning the parameter q', it is assumed as an epistemic uncertainty. Its variation range is chosen by a calibration with respect to the National Institute of Standards and Technology (NIST)[30] experimental data in terms of the saturation curve, considering the pressure as a function of temperature and the liquid volume as a function of temperature. The experimental uncertainties are available in the NIST Database and they are compared to several curves obtained with the method of Ref. 27 by varying q' (Figs. 4.4(a) and 4.4(b)). These figures show that the curves closer to the experimental data are the ones obtained by

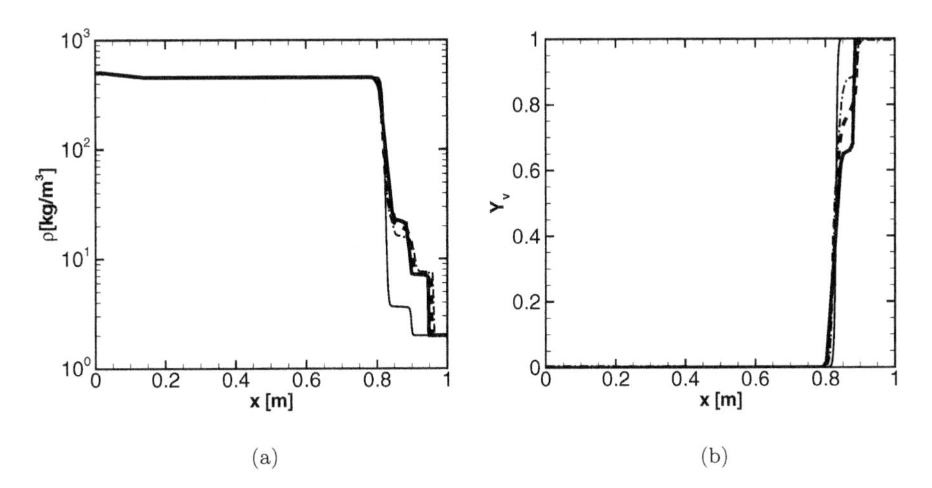

Fig. 4.3. (a) Mixture density at a time $t = 473$ μs: w/o heat and mass transfer (solid), the present model (dashed), Ref. 9 (dotted) and Ref. 2 (dashdot). (b) $Y_v = \alpha_1\rho_1/\rho$ at a time $t = 473$ μs: w/o heat and mass transfer (solid), the present model (dashed), Ref. 9 (dotted) and Ref. 2 (dashdot).

varying q' between -24470 and -24430. The left pressure and the temperature are considered as aleatory uncertainty, since they are directly related to the measurement uncertainties. A variation of 0.4% in terms of minimal/maximal value is then assumed relying on the literature.

The maximum and minimum levels of the numerical solutions (with respect to the uncertainties) of the mixture pressure and velocity at a time $t = 473$ μs are shown in Figs. 4.5(a) and 4.5(b), respectively. Variability of the different structures can be analysed. In particular, rarefaction wave seems to be quite not sensitive to the variation of input uncertainties. On the contrary, the position and intensity of the contact discontinuity vary much more, thus the prediction of these quantities is someway questionable. Also a strong influence on the evaporation front is observed.

In order to further analyse the robustness of the cavitation and the thermodynamic models, the Probability Density Function (PDF) is computed for several quantities of interest. In particular, the PDF of the pressure (at x=0.8) and of the velocity (at $x = 0.9$) are computed (see Figs. 4.6(a) and 4.6(b), respectively). Moreover, the cavitation length is computed, as $L_{cav} = x_2 - x_1$, where x_1 is the point for which $\alpha_g > 0$ and x_2 is the point for which $\alpha_g < 1$. Note that in this way, cavitation length is associated only to the variation of vapour fraction generated by the evaporation front.

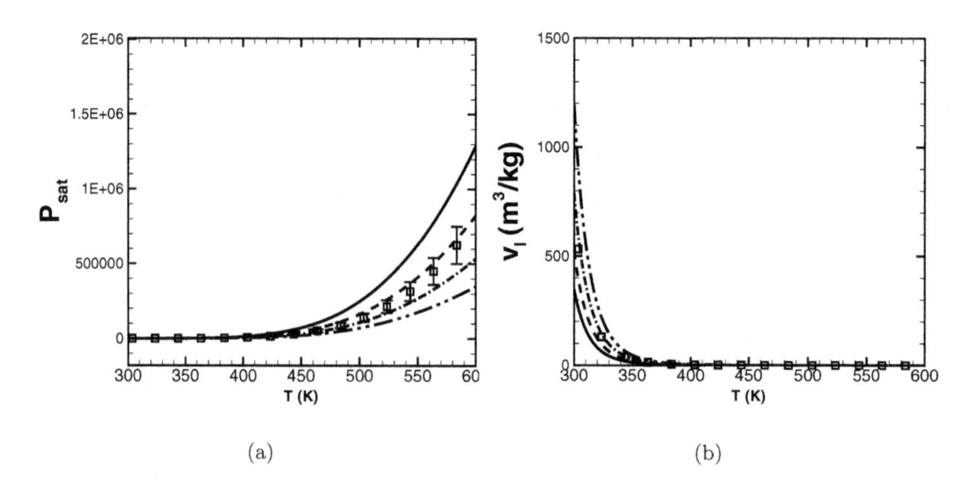

(a)

(b)

Fig. 4.4. Variability of the saturation curve (pressure-temperature) (a) w.r.t. q': $q' = -24430$ (solid), $q' = -24460$ (dashed), $q' = -24470$ (dashdot) and $q' = -24500$ (dashdot dotted). Variability of the saturation curve (liquid volume-temperature) (b) w.r.t. q': $q' = -24430$ (solid), $q' = -24460$ (dashed), $q' = -24470$ (dashdot) and $q' = -24500$ (dashdot dotted).

The PDF of the cavitation length is evaluated and reported in Fig. 4.7. As it can be observed, the most probable values lie between 0.06 m and 0.09 m. This behaviour makes questionable if the model is truly predictive in terms of cavitation length. A similar trend is observed also for the PDF of the velocity, between 240 m/s and 300 m/s (see Fig. 4.6(b)). Concerning the pressure (Fig. 4.6(a)), a different trend is observed since probability is concentrated in lower values of pressure, between 400 kPa and 600 kPa. These results clearly make arguable the quality of the transition prediction, showing anyway a low robustness with respect to the considered uncertainties.

6.2. *Two-phase expansion tube test*

Now, let us consider a tube of unit length, filled out with water at atmospheric pressure $P = 10^5$, density $\rho_2 = 1150$ kg/m^3 and temperature $T_2 = 354.728$ K. A small amount of vapour, $\alpha_1 = 10^{-2}$, is uniformly distributed in the whole domain (see Fig. 4.8). By assuming the flow in thermal and pressure initial equilibrium $T_2 = T_1$, the vapour density can be computed by combining Eqs. (4.16a) and (4.16b) and it is equal to $\rho_1 = 0.63$ kg/m^3. The solution is shown at a time $t = 3.2$ ms in Fig. 4.9.

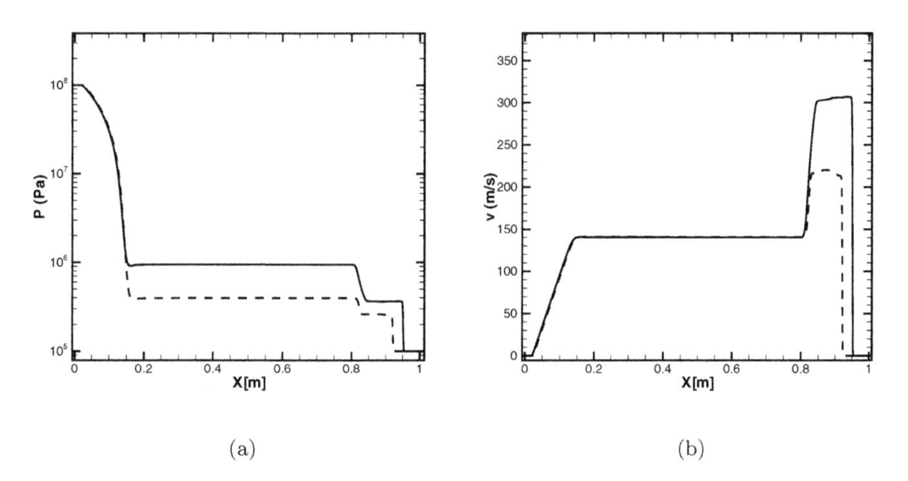

(a) (b)

Fig. 4.5. Maximum and minimum solutions of mixture pressure (a) and velocity (b) along the physical space, as a function of q'. For the pressure: solid line with $q' = -24469$, $P = 1.0039 \cdot 10^8$ and $T = 681.25$; dashed line with $q' = -24431$, $P = 1.00017 \cdot 10^8$ and $T = 681.25$. For the velocity: solid line with $q' = -24469$, $P = 1.0039 \cdot 10^8$ and $T = 681.25$; dashed line with $q' = -24431$, $P = 1.00017 \cdot 10^8$ and $T = 681.25$.

The computation is performed by using a mesh of 5000 cells. A velocity discontinuity ($v = -2$ m/s on the left and $v = 2$ m/s on the right) is set at x = 0.5 m.

Figure 4.9(a) shows the vapour volume fraction that increases in the center of the tube, where a mechanical expansion exists, and the pressure could attain the saturation value of about 0.5 bar (see Fig. 4.9(c)). By performing a comparison with the results obtained by Pelanti and Shyue[9] and by Zein *et al.*,[8] a very good agreement in terms of pressure and velocity (Figs 4.9(b) and 4.9(c)) can be observed. On the contrary, some differences can be observed in terms of vapour volume fraction (see Fig. 4.9(a)). Note that this difference could be explained by the different value used for the equation of state's parameter q'. In particular, here, q' is chosen in order to find the plateau pressure value reported in Refs. 8 and 9. This pressure value is equal to nearly 5×10^4 Pa in Fig. 4.9(c) and it corresponds to the saturation pressure related to the flow temperature in the tube.

In this case, three uncertainties are considered: the EOS parameter, q', the initial left velocity and the vapour volume fraction.

These parameters are the most influential in terms of variation of the solution. In particular, the initial velocity is the only thermo-mechanical

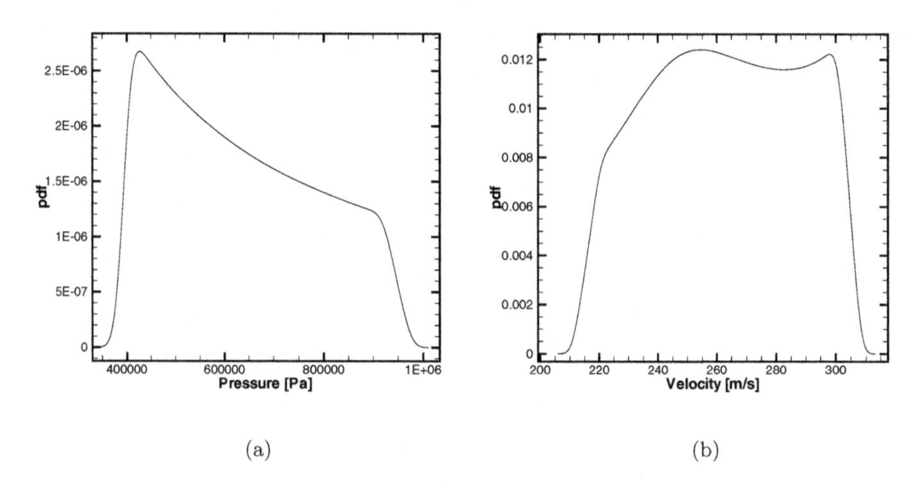

(a) (b)

Fig. 4.6. Probability density function of the pressure (a) at $x = 0.8$ and for the velocity (b) at $x = 0.9$.

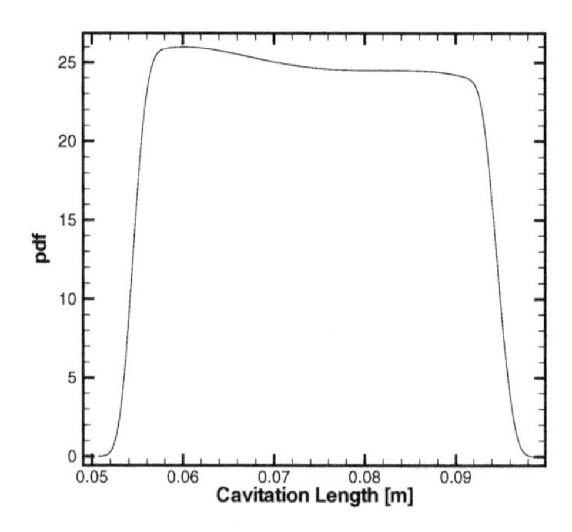

Fig. 4.7. Probability density function of the cavitation length.

parameter that determines the flow discontinuity in the tube and, consequently, the phase change. The initial vapour volume fraction represents the percentage of non-condensable gas. The presence of this gas has been observed also experimentally. So it is important to establish its influence in the computation. Finally, the parameter q' is an EOS parameter that

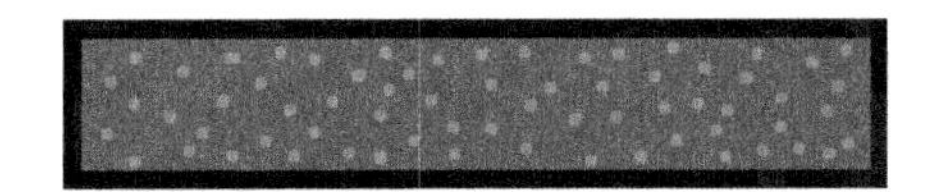

Fig. 4.8. Expansion tube geometry.

features a huge influence on pressure and velocity profile variations. The procedure for setting the variation range of the EOS parameter, q', is the same of the previous test case. Now, it provides a q' varying between -23500 and -23300. Figures 4.10(a) and 4.10(b) illustrate the variability of the saturation curve w.r.t. the parameter q'. The initial left velocity is assumed as an epistemic uncertainty, with a variation in terms of minimal/maximal values of $\pm 10\%$ with respect to the initial value of -2 m/s (treated with an uniform probability density function). A uniform distribution is chosen since this is the most conservative hypothesis if no additional knowledge is available on a given quantity. The variation of the initial vapour volume fraction is established by considering the existing literature about the existence of non-condensable gas in the liquid phase. In particular, relying on experimental observations for water at room temperature,[1] the initial gas volume fraction is assumed to vary between $5 \cdot 10^{-3}$ and $1 \cdot 10^{-2}$.

The maximum and minimum levels of the numerical solution (with respect to the uncertainties and at a time $t = 0.0032$ s) of the gas volume fraction, α_g, and of the pressure are computed and reported in Figs. 4.11(a) and 4.11(b), respectively. A large variability of the numerical solution is observed, thus showing the difficulty to use this test-case for cavitation model validation. Also for this test-case, ANOVA analysis is used for determining the most important uncertainties. The variable α_g seems to be dependent from all the uncertainties (uncertainties on q' varies from 35% to 58% along the axis, varies from 25% to 51% on the initial α_g, and from 13% to 36% on initial left velocity). On the contrary, the pressure variation in terms of level and wave position is mostly due to uncertainties on q' (from 25% to 45%) and on the initial α_g (from 55% to 60%).

Finally, the PDF of a quantity of interest is computed, i.e. in this case the pressure at $x = 0.8$ and at a time $t = 0.0032$ (reported in Fig. 4.12). The distribution of the pressure is nearly uniform except for two peaks that can be observed, which is mostly influenced by the initial left velocity (95% in terms of contribution to the variance) and by the initial gas fraction (5%).

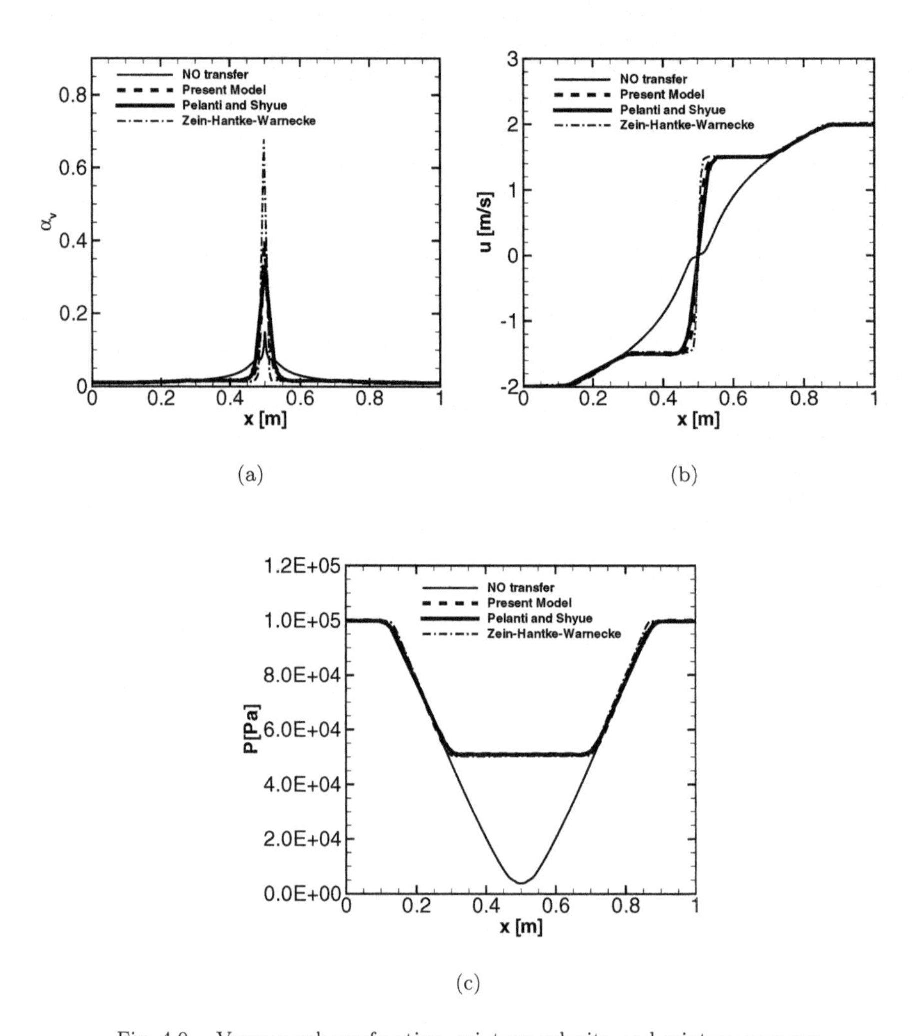

Fig. 4.9. Vapour volume fraction, mixture velocity and mixture pressure.

7. Conclusions and Perspectives

The present study is focused on the uncertainty propagation of some parameters of the cavitation model in order to estimate the variability of the cavitation-driven phenomena for metastable states. Stochastic analysis has shown that transition and thermodynamic modeling are extremely sensitive to some parameters. Some quantities of interest, such as the cavitation length or velocity at a specific location, feature a distribution that is nearly uniform in some specific ranges. This clearly has shown that the quality

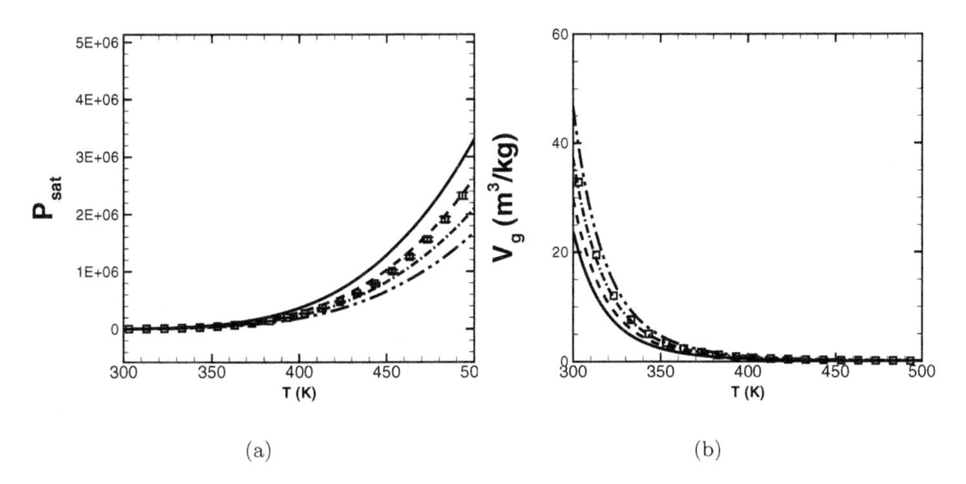

Fig. 4.10. Variability of the saturation curve (pressure-temperature) (a) w.r.t. q': $q' = -23200$ (solid), $q' = -23300$ (dashed), $q' = -23400$ (dashdot) and $q' = -23500$ (dashdot dotted). Variability of the saturation curve (liquid volume-temperature) (b) w.r.t. q': $q' = -23200$ (solid), $q' = -23300$ (dashed), $q' = -23400$ (dashdot) and $q' = -23500$ (dashdot dotted)

of the prediction (in terms of cavitation length and other quantities of interest) seem questionable considering the ranges of variation obtained by calibration with respect to the saturation curve.

Future actions will be focused on the validation of different classes of fluids, where uncertainties on thermodynamic conditions can be much more reduced.

References

1. C. Brennen, *Cavitation and Bubble Dynamics*. Oxford University Press (1995).
2. R. Saurel, F. Petitpas and R. Abgrall, Modelling phase transition in metastable liquids: application to cavitating and flashing flows, *J. Fluid Mech.* **607**, 313–350 (2008).
3. J. Simoes-Moreira and J. Shepherd, Evaporation waves in superheated dodecane, *J. Fluid Mech.* **382**, 63–86 (1999).
4. R. Saurel and O. LeMetayer, A multiphase model for compressible flows with interfaces, shocks, detonation waves and cavitation, *J. Fluid Mech.* **431**, 239–271 (2001).
5. R. Abgrall and M. G. Rodio, Discrete equation method (dem) for the simulation of viscous, compressible, two-phase flows, *Comput. Fluids* **91**(0),

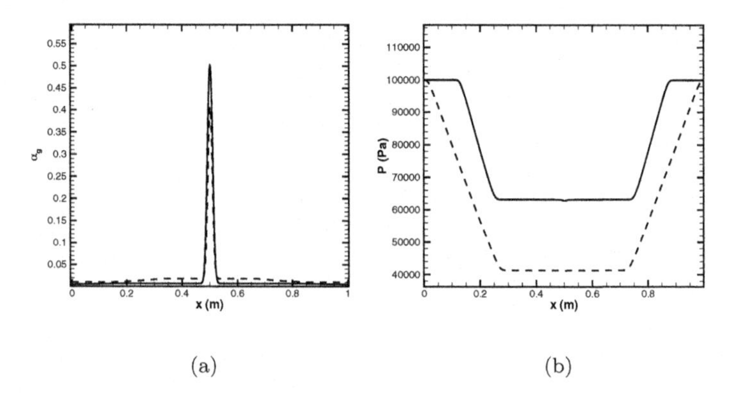

(a) (b)

Fig. 4.11. Maximum and minimum solutions of the gas volume fraction (a) and of the mixture pressure (b) along the physical space, as a function of q'. For the gas volume fraction: solid line with $q' = -23305$, $v = -2.19$ m/s and $\alpha = 1.23 \cdot 10^{-4}$; dashed line with $q' = -23495$, $v = -1.81$ m/s and $\alpha = 9.77 \cdot 10^{-4}$. For the mixture pressure: solid line with $q' = -23305$, $v = -1.81$ m/s and $\alpha = 9.77 \cdot 10^{-4}$; dashed line with $q' = -23495$, $v = -1.81$ m/s and $\alpha = 2.16 \cdot 10^{-4}$.

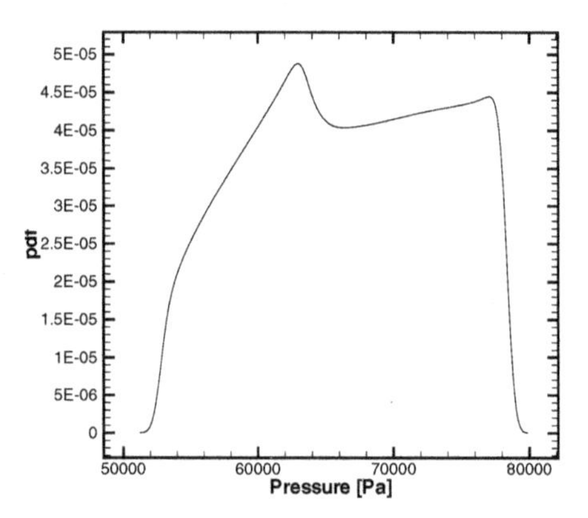

Fig. 4.12. Probability density function of the pressure at $x = 0.8$.

164–181 (2014).

6. A. Murrone and H. Guillard, A five equation reduced model for compressible two-phase flow problems, *J. Comput. Phys.* **202**, 664–698 (2005).
7. E. Goncalves, Modeling for non-isothermal cavitation using 4-equation models, *Int. J. Heat Mass Transf.* **76**, 247–262 (2014).

8. A. Zein, M. Hantke and G. Warnecke, Modeling phase transition for compressible two-phase flows applied to metastable liquids, *J. Comput. Phys.* **229**, 2964–2998 (2010).
9. M. Pelanti and K. M. Shyue, A mixture-energy-consistent six-equation two-phase numerical model for fluids with interfaces, cavitation and evaporation waves, *J. Comput. Phys.* **259**, 331–357 (2014).
10. F. Daude, P. Galon, Z. Gao and E. Blaud, Numerical experiments using a HLLC-type scheme with {ALE} formulation for compressible two-phase flows five-equation models with phase transition, *Comput. Fluids* **94**(0), 112–138 (2014).
11. Y. Wang, L. Qiu, R. D. Reitz and R. Diwakar, Simulating cavitating liquid jets using a compressible and equilibrium two-phase flow solver, *International J. Multiphase Flow* **63**(0), 52–67 (2014).
12. R. Abgrall, M. Rodio and P. Congedo. Towards an efficient simulation of cavitating flows with real gas effects and uncertainty quantification. In *WCCM XI-ECCM V-ECFD VI*, Barcelona, Spain, pp. 4779–4790 (2014).
13. G. Faccanoni, S. Kokh and G. Allaire, Approximation of liquid–vapor phase transition for compressible fluids with tabulated {EOS}, *C. R. Math.* **348** (7–8), 473–478 (2010).
14. M. G. Rodio and R. Abgrall, An innovative phase transition modeling for reproducing cavitation through a five-equation model and theoretical generalization to six and seven-equation models, *Int. J. Heat Mass Transf.* **89** 1–35 (2015).
15. S. Li, Z. G. Zuo and S. C. Li, Stochastic study of cavitation bubbles near boundary wall, *J. Hydrodynamics* **18**(3), 487–491 (2006).
16. S. J. Fariborza, D. G. Harlowa and T. J. Delpha, Intergranular creep cavitation with time-discrete stochastic nucleation, *Acta Metall.* **34**(7), 1433–1441 (1986).
17. E. Giannadakis, D. Papoulias, M. Gavaises, C. Arcoumanis, C. Soteriou, and W. Tang. Evaluation of the predictive capability of diesel nozzle cavitation models. In *Proceedings of SAE International Congress*, Detroit (2007).
18. S. K. Mishra, K. Sudib, P. A. Deymier, K. Muralidharan, G. Frantziskonis, S. Pannala, and S. Simunovic, Modeling the coupling of reaction kinetics and hydrodynamics in a collapsing cavity, *Ultrason Sonochem.* **17**(1), 258–265 (2010).
19. L. Wilczynski, Stochastic modeling of cavitation phenomena in turbulent flow, *Adv. Fluid Mech. III* **29**, 1–10 (2000).
20. T. Goel, S. Thakur, R. T. Haftka, W. Shyy and J. Zhao, Surrogate model-based strategy for cryogenic cavitation model validation and sensitivity evaluation, *Int. J. Numer. Methods Fluids* **58**, 969–1007 (2008).
21. C.-C. Tseng and L.-J. Wang, Investigations of empirical coefficients of cavitation and turbulence model through steady and unsteady turbulent cavitating flows, *Comput. Fluids* **103**, 262–274 (2014).
22. M. Rodio and P. Congedo, Robust analysis of cavitating flows in the Venturi tube, *Eur. J. Mech. B Fluids* **44**(0), 88–99 (2014).

23. R. Abgrall and R. Saurel, Discrete equations for physical and numerical compressible multiphase mixtures, *J. Comput. Phys.* **186**, 361–396 (2003).
24. R. Abgrall and V. Perrier, Asymptotic expansion of multiscale numerical scheme for compressible multiscale flow, *SIAM J. Sci. Comput.* **5**, 84–115 (2006).
25. M. G. Rodio, P. M. Congedo and R. Abgrall, Two-phase flow numerical simulation with real-gas effects and occurrence of rarefaction shock waves, *Eur. J. Mech. B Fluids* **45**, 20–35 (2014).
26. M. Rodio, P. Congedo and R. Abgrall, Two-phase flow numerical simulation with real-gas effects and occurrence of rarefaction shock waves, *Eur. J. Mech. B Fluids* **45**, 20–35 (2014).
27. O. Lemetayer, J. Massoni and R. Saurel, Elaboration des lois detat dun liquide et de savapeur pour les modeles decoulements diphasiques, *Int. J. Thermal Sci.* **43**, 265–276 (2003).
28. P. M. Congedo, C. Corre and J.-M. Martinez, Shape optimization of an airfoil in a BZT flow with multiple-source uncertainties, *Comput. Methods Appl. Mech. Engrg.* **200**(1), 216–232 (2011).
29. T. Crestaux, O. Le Maître and J.-M. Martinez, Polynomial chaos expansion for sensitivity analysis, *Reliab. Eng. Sys. Saf.* **94**(7), 1161–1172 (2009).
30. http://webbook.nist.gov/chemistry/fluid/.

Chapter 5

Uncertainty Quantification in Structural Engineering: Current Status and Computational Challenges

Sayan Gupta

Department of Applied Mechanics
Indian Institute of Technology Madras
Chennai 600036, India
sayan@iitm.ac.in

Debraj Ghosh

Department of Civil Engineering
Indian Institute of Science
Bangalore 560012, India
dghosh@civil.iisc.ernet.in

A review on the computational aspects of uncertainty quantification (UQ) as applicable to structural systems is presented. The review is limited to probabilistic frameworks for modelling of uncertainty and its propagation to the response. The topics reviewed include stochastic finite element, methods for analytical and Monte Carlo simulation based structural reliability analysis, time variant reliability, random fatigue, inverse methods in system identification, and parallelization. The chapter ends with identifying possible avenues for future research.

1. Introduction

Uncertainties in structural engineering systems arise due to the lack of precise knowledge in estimating the expected loadings on the structure during its lifetime, limitations in modelling the structure and material behaviour, availability of incomplete and/or noisy data sets from measurements, and unknown induced errors due to simplifying assumptions or limitations in analysis. Due to these uncertainties the behaviour of structural systems deviate from their designed performance. Therefore, quantifying the effects of these uncertainties is important in the context of estimating structural safety and evolving design procedures that meet specified safety levels.

The presence of uncertainties and their deleterious effects on the structural performance have led to the development of the *factor of safety* approach in design. Here, the design load is selected by multiplying the expected maximum load by a factor greater than unity, and the design capacity is estimated by dividing the structural capacity by a factor greater than unity. Selection of these factors is subjective, lacks scientific rigor, leads to suboptimal designs, and makes it impossible to estimate the structural reliability - a measure that is increasingly being demanded by regulatory bodies. Modern approaches to deal with uncertainties adopt probabilistic methodologies for quantifying the propagation of uncertainties into the response, and arrive at designs that meet target reliability levels. These approaches are collectively referred to as uncertainty quantification (UQ). UQ comprises broadly two components: uncertainty modelling and quantifying its propagation. The goal in modelling is to find the best approximations (or descriptors) of the uncertain quantities in the field equations, whereas in propagation the challenge lies in finding accurate and computationally economical methods to estimate the variability in the response quantity of interest.

The evolution of theories of UQ has been traditionally based on adopting probabilistic frameworks for quantifying the uncertainties associated with the problem. An alternative class of studies based on possibilistic frameworks has received impetus in the last decade.[1,2] The basic difference between the probabilistic and possibilistic approaches to structural reliability assessment is that in the former framework, attention is focussed on quantifying the propagation of uncertainties which have been modelled using classical theories of probability, while the latter approach is devoted to quantifying the ignorance associated with the uncertainties and how this ignorance propagates into the system response. Here, one uses concepts of interval algebra, fuzzy logic, info-gap models and other related theories. However, irrespective of the underlying principle, it is clear that both approaches require overcoming significant computational challenges for structural reliability estimation. The focus of this review will however be limited to the probabilistic framework.

The organization of this chapter is as follows: Section 2 provides discussions on modelling and methods for analysis of uncertainty propagation in systems. The usefulness of these methods for structural reliability assessment is presented in Sec. 3. Issues related to time variant reliability analysis that involves using theories of random processes are discussed in

Sec. 4. Problems related to system identification associated with health monitoring is discussed in Sec. 5. Section 6 focuses on parallel computation in UQ. The chapter concludes with discussions on existing challenges and possible future directions in Sec. 7.

2. Uncertainty Modelling and Propagation

A major thrust of current research in UQ is on developing methodologies for quantifying the propagation of the uncertainties into the response. The field equations, in their most general form, are partial differential equations, whose coefficients could be random variables and random processes. Non-homogenous terms are represented in terms of stochastic processes in time and space. Typical examples of properties modelled as random variables include spring stiffnesses, point mass, magnitude of point loads.[3] Random processes are used to model continuous parameteric functions; this includes elastic properties modelled as spatial random process,[4,5] earthquake and wind loads modelled as temporal processes[6–8] or as spatio-temporal processes.[9,10] Detailed discussions on the modelling aspects are available.[11,12] Solving the stochastic field equations is a mathematically and numerically challenging task. The discussions related to the solutions of such equations presented here are restricted to structural engineering applications.

In most structural engineering applications, the focus has been on obtaining weak form solutions for the field equations that enable quantifying the response quantities of interest in a probabilistic sense. This involves discretization of the response fields in terms of response variables using finite element based approaches. An additional difficulty in seeking weak form solutions involve discretization of the stochastic fields. This has been the focus of active research in the literature and is generally classified as stochastic finite element method (SFEM).[11,13–15] The following section provides a review of a special class of such methods that uses spectral bases for approximation.

2.1. *Spectral stochastic finite element methods*

Analogous to deterministic finite element methods for solving ordinary and partial differential equations (ODEs and PDEs), stochastic PDEs can be solved efficiently using SFEM. These methods can be perturbation based,[16] or based on approximation in a functional space constructed by global bases such as polynomials — also referred to as the spectral stochastic finite element methods (SSFEM).[17] The first step in SSFEM is to discretize the random fields. This discretization is necessary to reduce a random field — which is an infinite dimensional object — to a finite dimensional approximation. Several discretization methods have been developed, such as local averaging,[18,19] midpoint,[20,21] weighted integral,[22] optimal linear expansion.[23,24] Series representations such as Karhunen-Loève (KL)[25] and polynomial chaos expansions (PCE)[17] have been popular choices in recent years and shown to be optimal.[23] Unification of some of these methods has also been attempted.[26] PCE is also used in response approximation, as will be described later in this section.

For a random process $\kappa(\boldsymbol{x})$ with known mean $\bar{\kappa}(\boldsymbol{x})$ and covariance function $C(\boldsymbol{x}_1, \boldsymbol{x}_2)$, the KL expansion, which is a mean-square convergent representation, is

$$\kappa(\boldsymbol{x}) = \bar{\kappa}(\boldsymbol{x}) + \sum_{i=1}^{\infty} \sqrt{\lambda_i}\phi_i(\boldsymbol{x})\xi_i, \tag{5.1}$$

where λ_i and $\phi_i(\boldsymbol{x})$ are the ith eigenvalue and the corresponding eigenfunction of the covariance kernel, respectively, and ξ_i-s are uncorrelated random variables with zero mean and unit variance. Note that all random variables and processes considered here are assumed to be square-integrable. The integral eigenvalue problem (IEVP) of the covariance kernel is of the form

$$\int_{\mathcal{D}} C(\boldsymbol{x}_1, \boldsymbol{x}_2)\phi_i(\boldsymbol{x}_1)d\boldsymbol{x}_1 = \lambda_i\phi_i(\boldsymbol{x}_2), \tag{5.2}$$

where $\mathcal{D}$ denotes the physical domain over which the process is defined. Symmetry and positive semidefiniteness of the kernel ensure that the eigenvalues are real and non-negative. A truncated version of the KL expansion,

$$\hat{\kappa}(\boldsymbol{x}) \equiv \bar{\kappa}(\boldsymbol{x}) + \sum_{i=1}^{L} \sqrt{\lambda_i}\phi_i(\boldsymbol{x})\xi_i, \tag{5.3}$$

is used in practical computations. The eigenvalues are indexed in descending order. As the correlation length of the process increases, the eigenvalues decay faster, and the truncation level L reduces. A numerical study on convergence of the KL expansion is available.[27] Covariance kernels with exponential decay and variants are commonly used. It is also important to ensure that the kernel is differentiable everywhere, a criterion that may require modification of existing kernels.[28] For Gaussian processes, selection of the random bases ξ_i is straightforward — they are independent standard normal variables. However, some non-Gaussian processes (such as lognormal and beta) also can be modelled using KL expansion.[29–31]

The IEVP in Eq. (5.2) can be solved analytically for a limited class of kernels, and only over simple domains — an interval in one dimensions or over a box in two or three dimensions.[25] For all other cases a numerical method needs to be invoked such as Galerkin finite element (FE)[17] or quadrature.[32] The Nyström method is effective for using the quadrature. For FE, usually the mesh used for solving the mechanics problems is reused. In several studies[20,26,33] it was found that this mesh selection is not optimal. This optimality issue also appears for other discretizations.[23,34] To address this issue, it was proved recently[35] that the KL approximation is independent of the size or shape of the domain. Following this observation, a method was proposed there that significantly reduced the computational cost. There have been a number of reported methods for computation acceleration such as adaptive FE,[36] hierarchical sparse matrix approximation,[37] fast multipole method,[38] and finite cell method.[39] Modifications in the bases have also been proposed such as trigonometric basis set[40] and Fourier-KL.[41]

The uncertainty propagation is explained here using the stochastic elliptic equations

$$\begin{aligned} \mathcal{L}(u(\boldsymbol{x},\boldsymbol{\xi})) &= f(\boldsymbol{x},\boldsymbol{\xi}), \quad \boldsymbol{x} \in \mathcal{D}, \\ u(\boldsymbol{x},\boldsymbol{\xi}) &= u_b(\boldsymbol{x},\boldsymbol{\xi}), \quad \boldsymbol{x} \in \partial\mathcal{D}, \end{aligned} \tag{5.4}$$

holding almost everywhere on Ω; here $\mathcal{D}$ and $\partial\mathcal{D}$ denote the physical domain and its boundary, respectively, and $\mathcal{L}(\cdot)$ is an operator. Let the joint probability density function (pdf) of $\boldsymbol{\xi}$ be denoted by $p(\boldsymbol{\xi})$. The solution to Eq. (5.4) is expressed in an orthogonal polynomial expansion as

$$u(\boldsymbol{x},\boldsymbol{\xi}) = \sum_{i=0}^{\infty} u_i(\boldsymbol{x})\psi_i(\boldsymbol{\xi}); \quad u_i(\boldsymbol{x}) \in \mathbb{R}, \tag{5.5}$$

where the orthogonal random polynomials $\psi_i(\boldsymbol{\xi})$ serve as a basis set of a Hilbert space, satisfying the relations

$$\begin{aligned} \psi_0 \equiv 1\,, \quad \mathbb{E}\{\psi_i(\boldsymbol{\xi})\} &= 0 \ \text{ for } \ i > 0, \\ \mathbb{E}\{\psi_i(\boldsymbol{\xi})\psi_j(\boldsymbol{\xi})\} &= \delta_{ij}\mathbb{E}\{\psi_i^2(\boldsymbol{\xi})\} \end{aligned} \tag{5.6}$$

and $u_i(\boldsymbol{x})$ are the unknown projections. Here $\mathbb{E}\{\cdot\}$ is the expectation operator, such that

$$\mathbb{E}\{g(\boldsymbol{\xi})\} = \int_{\mathbb{R}^d} g(\boldsymbol{\xi})p(\boldsymbol{\xi})d\boldsymbol{\xi}. \tag{5.7}$$

When $\boldsymbol{\xi}$ constitute a vector of standard normal variables, these polynomials are Hermite, and the expansion is referred to as the PCE. For non-Gaussian cases, the Askey scheme of polynomials follows, terming the expansion as generalized PCE (gPCE).[42,43] Further numerical generalization of this expansion, termed as arbitrary chaos,[44] has also been developed. Again for practical computation the polynomial expansion is truncated as

$$\hat{u}(\boldsymbol{x}, \boldsymbol{\xi}) = \sum_{i=0}^{P-1} u_i(\boldsymbol{x}) \psi_i(\boldsymbol{\xi}); \qquad u_i(\boldsymbol{x}) \in \mathbb{R}, \tag{5.8}$$

where P depends upon d — the dimension of $\boldsymbol{\xi}$, and the highest degree of the polynomial bases. The next task is to find the chaos coefficients $u_i(\boldsymbol{x})$. Let the deterministic FE discretization (spatial) performed to solve the mechanics problem lead to n unconstrained degrees of freedom (DOF). Accordingly, Eq. (5.8) now takes the form

$$\hat{\boldsymbol{u}}(\boldsymbol{\xi}) = \sum_{i=0}^{P-1} \boldsymbol{u}_i \psi_i(\boldsymbol{\xi}); \qquad \hat{\boldsymbol{u}}, \boldsymbol{u}_i \in \mathbb{R}^n. \tag{5.9}$$

The unknown coefficients $\{\boldsymbol{u}_i\}_{i=0}^{P-1}$ can be estimated two ways. The first way is through the strong formulation - also referred to as the non-intrusive method. Accordingly, the orthogonality of the chaos bases is directly used in the untruncated version of Eq. (5.9), leading to the estimate

$$\boldsymbol{u}_i = \frac{\mathbb{E}\{\boldsymbol{u}(\boldsymbol{\xi})\psi_i(\boldsymbol{\xi})\}}{\mathbb{E}\{\psi_i^2(\boldsymbol{\xi})\}}, \qquad i = 0, \ldots, P-1. \tag{5.10}$$

The denominator in this expression is readily available in a tabular form, and the numerator is evaluated using statistical simulations. This statistical simulation is expensive, and to reduce this cost the second way — the weak formulation, or intrusive method — may be used. According to this method, Eq. (5.9) is substituted in spatially discretized form of the governing equation (5.4) and a Bubnov-Galerkin projection is performed on the chaos bases. These steps yield an nP-dimensional system of linear algebraic equations, which are then solved using standard numerical methods.

SSFEM has been successfully used in solving a variety of problems, such as random eigenvalue problems,[45–47] response prediction for linearly vibrating stochastic systems,[48] nonlinear dynamical systems,[17] fluid-structure interaction problems,[44,49] random fatigue characterization,[50,51] and even for solving inverse problems.[52,53] There have been a number of further developments on increasing the accuracy and speeding up the computation such as sparse chaos expansion,[54] sparse grid or Smolyak cubature,[55,56] stochastic collocation,[57] enrichment of the basis set,[58] multielement gPCE,[59] and hybrid methods.[60] Recent reviews on SSFEM are available in.[33,61,62]

3. Structural Reliability

3.1. *Analytical methods*

A first step in reliability assessment lies in modelling the uncertain loads — which can have spatio-temporal randomness — and in modelling the structural system. The response of a structural system, $\boldsymbol{Y}$ can be represented in the general form as $\boldsymbol{Y} = \boldsymbol{g}(\boldsymbol{X}(\gamma_1); \gamma_2)$, where, $\boldsymbol{X}(\gamma_1)$ represents the uncertain excitations that are equivalently represented in terms of a vector of random variables γ_1, $\boldsymbol{g}(\cdot)$ represents the functional form that provides a relation between the input forces and the structure response and depends on the mathematical model while γ_2 represents a vector of mutually correlated random variables that represents the uncertainties associated with the material properties of the structure model. Clearly, the response $\boldsymbol{Y} \equiv \boldsymbol{Y}(\boldsymbol{\theta})$ where, $\boldsymbol{\theta} = [\gamma_1 \; \gamma_2]^T$ denotes the vector of random variables that forms the support of $\boldsymbol{Y}(\boldsymbol{\theta})$. The failure probability P_f is mathematically defined as

$$P_f = \int_{h(\boldsymbol{\theta}) \leq 0} p_{\boldsymbol{\theta}}(\boldsymbol{s}) \, \mathrm{d}\boldsymbol{s}, \tag{5.11}$$

where, $h(\boldsymbol{\theta}) \leq 0$ indicates the failure domain defined in the $\boldsymbol{\theta}$-space and $p_{\boldsymbol{\theta}}(\boldsymbol{s})$ is the joint pdf of $\boldsymbol{\theta}$. The integral in Eq. (5.11) is of dimension equal to that of the vector $\boldsymbol{\theta}$. The difficulties in evaluating the integral in Eq. (5.11) are: (a) the dimension of the integral may be large, (b) usually, the failure domain defined by $h(\boldsymbol{\theta}) \leq 0$ is not explicitly defined, (c) even if $h(\boldsymbol{\theta}) \leq 0$ is available in a functional form, its behaviour may not be smooth, (d) knowledge about the joint pdf $p_{\boldsymbol{\theta}}(\boldsymbol{s})$ is usually not available, and (e) the evaluation of the function $h(\boldsymbol{\theta})$ may not be easy. A comprehensive review of the methods for reliability analyses that develops approximations by which these difficulties are addressed is available,[12] and is not repeated here. Instead, after a very brief discussion on the methodologies available for evaluating Eq. (5.11), this review will primarily focus on the computational challenges in time variant reliability analysis.

When $\boldsymbol{\theta}$ constitutes a vector of standard normal random variables and $h(\boldsymbol{\theta})$ is a linear function of $\boldsymbol{\theta}$ of the form $h(\boldsymbol{\theta}) = a_0 + \sum a_i \theta_i$, it can be shown that Eq. (5.11) can be evaluated exactly with $P_f = \Phi(-\beta)$, where, $\Phi(\cdot)$ is the standard Gaussian distribution function and β is a parameter that depends on the coefficients a_0 and a_i. The ability to represent the failure probability and all the related uncertainties associated with the system through a single parameter β constitutes a very important development in structural reliability theories. Of the several definitions for β that are available in the literature, the Hasofer-Lind reliability index,[63] denoted by β_{HL}, is the most widely used. A geometrical interpretation of β_{HL} is that it represents the shortest distance of the linear function $h(\boldsymbol{\theta})$ from the origin in the multi-dimensional $\boldsymbol{\theta}$-space; see Fig. 5.1. Qualitatively, β_{HL} enables

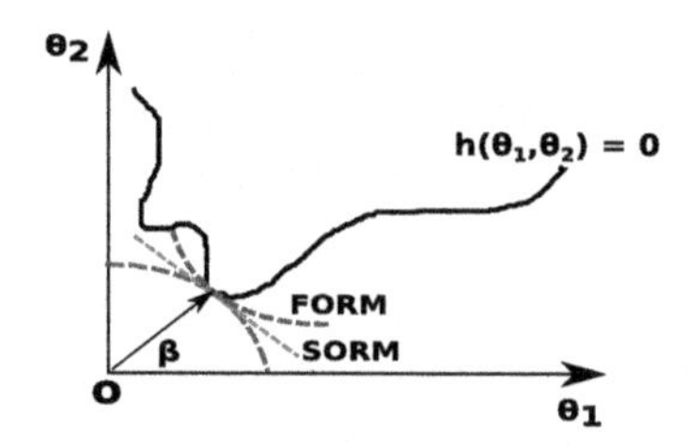

Fig. 5.1. Schematic diagram for first passage failure.

comparisons of the relative safety of similar structural systems.

In reality, in most structural engineering applications, $\boldsymbol{\theta}$ is neither standard normal, nor is $h(\boldsymbol{\theta})$ linear. In such situations, a two-step procedure is adopted: (a) first, the problem is transformed into the standard normal space and (b) subsequently, an optimization algorithm[64] is used to find the shortest distance from the origin to the nonlinearly transformed failure surface. These algorithms are based on the principles of Taylor series expansion of $h(\boldsymbol{\theta})$. Methods based on first order Taylor series expansion - also known as first order reliability methods or FORM, have been shown to lead to inaccurate estimates of P_f unless $h(\boldsymbol{\theta})$ satisfies specific functional forms. This has led to the development of methods which consider higher order terms in the Taylor series expansions of the failure surface giving rise to second-order reliability methods (SORM)[65–71] and higher-order reliability methods.[72–74] For more discussions on FORM, SORM, and their variants, the reader is directed to the review[12] .

3.2. *Monte Carlo simulation based approaches*

FORM, SORM, or any other variant of such methods are essentially approximate in nature and lead to reliability indices which provide a quantitative measure of the relative reliability of different structural systems. The usefulness of these methods are somewhat limited if one is interested in estimating the failure probabilities in the absolute sense. These methods are also handicapped by the lack of exactness in dealing with structural nonlinearities and parametric excitations. These difficulties can be overcome through the use of Monte Carlo simulation (MCS) based approaches. MCS based approaches are suitable for reliability estimation for any problem amenable for deterministic numerical analysis and is applicable to situations where it is possible to digitally generate ensemble of the uncertainties — structural or loadings — that specify prescribed probability laws.[75] The primary steps in MCS involve simulating an ensemble of realizations for $\boldsymbol{\theta}$ that conform to the prescribed probability laws $p_{\boldsymbol{\theta}}(\boldsymbol{\theta})$, and subsequently, evaluating the function $h(\boldsymbol{\theta})$ for each realization. An estimate of P_f is obtained as

$$\hat{P}_f = \frac{1}{N} \sum_{i=1}^{N} I[h_i(\boldsymbol{\theta}) \leq 0], \tag{5.12}$$

where N is the sample size, $h_i(\boldsymbol{\theta})$ is the ith realization of the evaluation of the performance function, and $I[\cdot]$ is an indicator function that takes value of unity if $h_i(\boldsymbol{\theta}) \leq 0$ and zero otherwise. It can be shown that the estimator in Eq. (5.12) is unbiased and has a minimum variance given by $\sigma^2 = \hat{P}_f(1 - \hat{P}_f)/N$. It follows that the accuracy of the estimator can be improved by increasing N for a given level of P_f; a thumb rule being that $N \approx O(10/P_f)$. It is therefore clear that to obtain accurate estimates of P_f, one needs to solve a large number of deterministic mechanics problems, each of which may require intensive computational effort. This has led to the development of several specialized sampling techniques that enable estimating low order failure probabilities without having to resort to prohibitively large sample sizes. These methods are classified as variance reduction techniques (VRT) in MCS.[76,77] The most popular of the VRT is importance sampling and its variants. The basic principle of this method lies in sampling from a pdf $w(\boldsymbol{\theta})$ such that the number of realizations that lie in the failure domain are much higher than if the sampling was carried out from the parent pdf $p_{\boldsymbol{\theta}}(\boldsymbol{\theta})$.[78] The failure probability is expressed as

$$P_f = \int_{-\infty}^{\infty} \frac{I[h(\boldsymbol{\theta}) \leq 0] p_{\boldsymbol{\theta}}(\boldsymbol{\theta})}{w(\boldsymbol{\theta})} w(\boldsymbol{\theta}) \, \mathrm{d}\boldsymbol{\theta}, \tag{5.13}$$

where the realizations lying in the failure domain are weighted based on the ratio $p_{\boldsymbol{\theta}}(\boldsymbol{\theta})/w(\boldsymbol{\theta})$. It can be shown that the estimator $\hat{P}_f$, given by

$$\hat{P}_f = \frac{1}{N} \sum_{i=1}^{N} \frac{I[h(\theta_i) < 0] p_{\boldsymbol{\theta}}(\theta_i)}{w(\theta_i)}, \tag{5.14}$$

is unbiased with minimum variance given by

$$\mathrm{Var}[\hat{P}_f] = \frac{1}{N-1}\left[\frac{1}{N}\sum_{i=1}^{N}\left\{\frac{I[h(\theta_i) \leq 0]p_{\boldsymbol{\theta}}(\theta_i)}{w(\theta_i)}\right\}^2 - \hat{P}_f^2\right]. \tag{5.15}$$

Clearly, a crucial step in the implementation of importance sampling lies in the selection of the importance sampling pdf $w(\boldsymbol{\theta})$. A review of methods for selection of $w(\boldsymbol{\theta})$ in the context of structural reliability calculations and their implementation is available.[12] An adaptation of importance sampling is the stratified sampling method, which is applicable when the domain of integration $\{\Omega : h(\boldsymbol{\theta}) \leq 0\}$ is subdivided into a set of disjoint domains $\Omega_1, \ldots, \Omega_k$, and estimates of the contribution to the integral from each of these subdomains are obtained using importance sampling. The overall estimate of the integral is obtained as $\hat{P}_f = \sum_{j=1}^{k} \hat{P}_{f_j}$, where, $\hat{P}_{f_j}$ is the estimate of the integral over the subdomain Ω_j. The variance of the estimator is given by $\mathrm{Var}[\hat{P}_f] = \sigma_1^2/m_1 + \cdots + \sigma_k^2/m_k$, where, σ_j^2 is the variance in Ω_j and m_j is the sample size used in estimating $\hat{P}_{f_j}$, and is less than σ^2/m, if σ_j^2 are more or less of the same order and $m = m_1 + \cdots + m_k$. Thus, this method is useful if the domain can be subdivided such that σ_j^2 are more or less equal in all subdomains. Several other forms of VRT have been discussed in the literature; see for example, the conditional expectation method, the generalized conditional expectation method, antithetic variate method, control variate technique and the method of latin hypercube sampling. A comparison of the performance of these VRT has been presented[79] with respect to failure probability calculations in a structural engineering problem.

4. Time-Variant Reliability

The problem of time variant reliability analysis in structural systems is fundamentally much more challenging. The response of a vibrating structural system, subjected to random loadings, is a random process in time. Failures in such structural systems are expressed either in terms of first passage failures, that is, a failure is deemed to occur if the response exceeds a specified threshold level at any instant of time within the structure's lifetime - or in terms of gradual accumulation of damage due to fatigue effects. In the latter case, when the loadings are random, the growth in fatigue damage is a random process. The first part of this review addresses the developments in the literature in establishing first passage failures. Subsequently, questions related to characterizing random fatigue damage is reviewed.

4.1. *First passage failures*

Mathematically, the probability of failure against first passage times in time variant problems is expressed as

$$P_f = 1 - P[X(t) > \alpha;\ \forall t \in (t_0, t_0 + T)], \tag{5.16}$$

where $X(t)$ is the structure response modelled as a random process in time t, α is the safe threshold, the exceedance of which at any instant of time is assumed to lead to failure, t_0 is the initial time, T is the duration of the design life, and $P[\cdot]$ is the probability measure; see Fig. 5.2 for a schematic. The evaluation of the probability in Eq. (5.16) requires knowledge of the

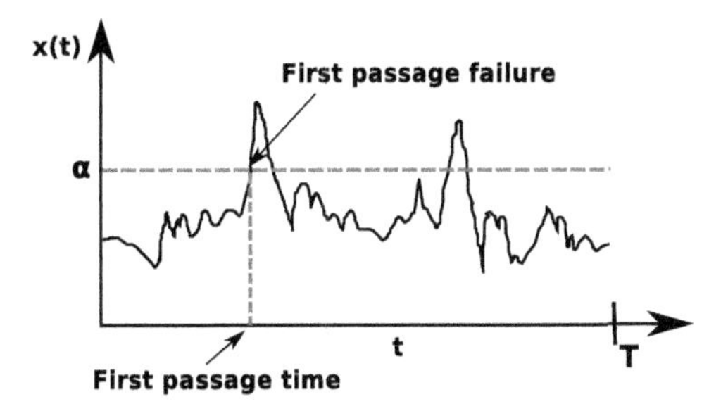

Fig. 5.2. Schematic diagram for first passage failure.

infinite dimensional joint pdf of the process $X(t)$ for all instants of time and its evaluation. Clearly, this is an infeasible task. This difficulty is bypassed by reframing the problem in a time invariant format. Defining a random variable

$$X_m = \max_{t_0 < t \leq t_0 + T} X(t), \tag{5.17}$$

where, X_m is the extreme value of the process $X(t)$ in time duration T, Eq. (5.16) can be recast in the equivalent form

$$P_f = P[X_m \leq \alpha] = P_{X_m}(\alpha). \tag{5.18}$$

Thus, the focus now shifts in estimating the extreme value distribution (EVD) of $X(t)$. The problem of extreme value distributions has received wide research attention in the literature.[80–83] The focus of early studies in the field have been on developing asymptotic forms of EVD for a sequence of independent and identically distributed (i.i.d.) random variables. It has been shown that for a scalar sequence of i.i.d. random variables, there exists only three asymptotic distributions — Gumbel, Weibull, and Fréchet distributions, and which are represented through the generalized EVD of the form

$$P[X_m \leq x] = \exp\left[-\left\{1+\xi\left(\frac{x-\mu}{\sigma}\right)\right\}^{-\frac{1}{\xi}}\right], \tag{5.19}$$

such that, $\xi(\frac{x-\mu}{\sigma}) > 0$, $\sigma > 0$ and $-\infty < \xi < \infty$. Here, μ, σ and ξ are respectively the location, scale, and shape parameters. Methods for estimating the domains of attraction for these asymptotic distributions are available.[82,83] Alternative methods based on the generalized Pareto distributions are also discussed in the literature.[83–85]

Random processes with finite correlation lengths can be generalized as a sequence of mutually dependent random variables, and hence, the results related to i.i.d. sequence of variables are not applicable. The problem of establishing EVD for mutually dependent random variables is a difficult problem as no unique limiting distributions can be generalized. It has been shown however, that under certain conditions of the dependence structure, a sequence of random variables have identical limit distributions as those of i.i.d. variables.[81] However, slight variations on the dependence structure can lead to dramatically different models for EVD.[83] Thus, alternative approaches for obtaining the EVD for random processes have been investigated.

A commonly used approach in estimating the EVD for random processes in structural reliability literature has been to use the outcrossing statistics, based on the assumption that level crossings can be modelled as a Poisson counting process. Thus, it can be shown that[86]

$$P_f \leq P_f(0) + [1 - P_f(0)]\left\{1 - \exp\left[-\int_{t_0}^{t_0+T} \nu(\tau)\mathrm{d}\tau\right]\right\}, \tag{5.20}$$

where $P_f(0)$ is the failure probability at $t = t_0$ and $\nu(\tau)$ is the mean outcrossing rate of $X(t)$ across the threshold α. Typically, it is assumed that $t_0 = 0$ and $P_f(0) \approx 0$. The mean outcrossing rate can be computed exactly using Rice's integral,[87,88] given by

$$\nu^+(\alpha;t) = \int_0^\infty \dot{x} p_{X(0)\dot{X}(0)}(\alpha, \dot{x}; t)\mathrm{d}\dot{x}. \tag{5.21}$$

Here $p_{X(0)\dot{X}(0)}(\cdot,\cdot)$ is the joint pdf of the process $X(t)$ and its instantaneous time derivative $\dot{X}(t)$. The crux in evaluating Rice's integral lies in estimating this joint pdf. A closed form analytical solution is available only when $X(t)$ is Gaussian as the joint pdf is completely characterized in terms of the mean and the standard deviation of $X(t)$ and $\dot{X}(t)$. However, in general, especially if the response of a structural system is obtained as a nonlinearly filtered process of the input process which may or may not be Gaussian, the marginal pdf of $X(t)$ may not be easy to estimate; the joint pdf $p_{X(0)\dot{X}(0)}(\cdot,\cdot)$ is usually therefore not available. Modelling such processes as Gaussian and performing reliability analysis lead to underestimation of the failure probability.[89]

Early studies involving crossings on non-Gaussian processes used linearization based procedures to obtain approximations on the bounds for the exceedance probabilities.[90] An alternative approach to studying the crossings of non-Gaussian processes is to use a class of transformed Gaussian processes.[91,92] The underlying principle of these approaches is to consider non-Gaussian processes $Z(t)$ which can be expressed as $Z(t) = g(X(t))$, where, $X(t)$ is a Gaussian process and $g(\cdot)$ is a continuous and increasing function. Thus, the crossings of $Z(t)$ can be approximated by studying the crossings of $X(t)$. The crux in this approach lies in selecting the function $g(\cdot)$ for which several methods have been proposed.[93–96] The drawback in using these class of models is the inability to include any additional information such as higher order statistics that might be available. These shortcomings are bypassed in studies that model the non-Gaussian processes as Laplace moving averaged (LMA) processes.[89,97,98] LMA processes are characterized by the mean, the spectrum and two additional parameters which can be the skewness and kurtosis and thus enables retaining the non-Gaussian characteristics.

An alternative method for estimating the joint pdf $p_{X(0)\dot{X}(0)}(x, \dot{x})$ is to solve the associated Fokker-Planck-Kolmogorov (FPK) equation for the dynamical system. The FPK equation is a partial differential equation for the transitional pdf of the state variables, when the equations of motion are expressed as stochastic differential equations.[99,100] The dimensions of the state-space equations are augmented by rewriting the band limited excitations as filtered white noise processes. The equations of motion are therefore in the form of a set of coupled stochastic differential equations (SDE) and have Markovian properties. While exact solutions for FPK equations are available only for a selected class of oscillators,[101,102] approximations can be obtained for a wider class of oscillators using methods such as equivalent linearization,[103] Gaussian-closure techniques,[104,105] stochastic averaging,[106,107] equivalent nonlinearization,[108] maximum entropy based methods,[109] and non-Gaussian closure based methods.[110–112] In recent years, with the emergence of inexpensive computing facilities, the use of numerical methods for the solution of the FP equation has gained wider acceptance. This has led to the development of numerical methods based on finite elements (FE),[113–116] finite difference (FD),[115] cell mapping techniques,[117] and path integrals (PI)[118–120] for solving the FP equation. However, extending these methods for the solution of FPK equations for higher order systems poses formidable challenges which appear to be insurmountable at present.[121]

A number of non-Gaussian processes that have been widely studied in the literature are those obtained as quadratic transformations of Gaussian processes. Many real loads, such as ocean waves and wind loadings, show considerable non-Gaussian features and are often modelled by second order Volterra series expansions with Gaussian inputs. Approximations for the mean outcrossing rate for second order stochastic Volterra series were obtained by rewriting the joint pdf $p_{X(0)\dot{X}(0)}(x, \dot{x})$ in terms of the joint characteristic function from the knowledge of the joint moments of $X(t)$ and $\dot{X}(t)$ and evaluating the Rice's integral using saddle point approximations.[122–124] The higher order joint moments of $X(t)$ and $\dot{X}(t)$ were approximated using methods based on Gram-Charlier series expansions, Edgeworth series, and maximum entropy principles.[124,125] A geometric approach for developing analytical approximations for the crossings of similar processes have been obtained by transforming the problem into polar coordinates and using the property that, for Gaussian processes[126] a zero correlation implies independence.

The problem of crossings of second order Volterra series expansions is in principle similar to non-Gaussian processes obtained as quadratic transformations of vector Gaussian processes, if one uses the Kac-Siegert representation for decomposing the second order process.[127] The problem of outcrossings of vector random processes have been studied as a problem in load combinations[128–132] where the outcrossings have been studied for a scalar process obtained as a summation of the component processes. Studies on multivariate extreme value distributions, which are important in the context of time variant reliability of structural systems, have however not received significant attention in the literature. A geometrical approach to approximate the EVD of Gaussian/non-Gaussian processes has been adopted.[133,134] An alternative approach based on approximating the crossings of a vector of mutually correlated processes as correlated Poisson point processes have been shown to lead to elegant formulation for the multivariate EVD.[135–137]

4.2. *Random fatigue*

Characterizing the random fatigue damage in vibrating structures subjected to random dynamic loadings constitutes an important step in reliability assessment and residual life estimation in ageing or existing structures. Computing the fatigue damage from a random time history typically involves (a) expressing the time history into an equivalent number of equivalent cycles, (b) estimating the incremental fatigue damage due to each such cycle from the Wohler curves and (c) applying a suitable damage accumulation rule to compute the total fatigue damage. For complete characterization of the random fatigue damage, one would need to estimate the pdf of the fatigue damage, at time t, which is essentially a random variable. However, this problem is mathematically very challenging, as this involves characterizing interval crossing statistics. Instead, the focus in the literature has been on estimating the bounds for the mean fatigue damage.

Computation of the mean fatigue damage can be carried out either in the time domain or frequency domain. Time domain analyses involve repeated fatigue analysis on an ensemble of time histories having identical probabilistic characteristics, and can be computationally expensive. On the other hand, frequency domain approaches enable fast and elegant methods for estimating the mean fatigue damage. A key feature in fatigue estimation lies in extracting the equivalent cycles from a random time history.[138] Several algorithms based on peak counting, level crossings counting, range counting, and rain-flow counting have been proposed.[139–142] Of these, the rain-flow cycle (rfc) counting algorithm[143] is accepted to be the most accurate. An equivalent but more suitable definition for the algorithm and amenable for mathematical treatment and statistical analysis has been developed.[144–147] Here, the expected fatigue damage is expressed as

$$\begin{aligned} \mathrm{E}[D^{rfc}(t)] &= \int_{-\infty}^{\infty}\int_{-\infty}^{v} \frac{\partial^2 f}{\partial u \partial v}(u,v)\mathrm{E}[N(u,v)]\mathrm{d}u\mathrm{d}v \\ &\quad + \int_{-\infty}^{\infty} \frac{\partial f}{\partial u}(u,u)\mathrm{E}[N(u)]\mathrm{d}u, \end{aligned} \tag{5.22}$$

where, $f(u,v)$ is the functional form of the damage accumulation rule, $\mathrm{E}[N(u,v)]$ is the expected interval crossings, and $\mathrm{E}[N(u)]$ is expected level crossings. While the expected level crossings can be computed from the Rice's formula, estimating the interval crossings is more difficult. However, a bound on the expected fatigue damage can be obtained from the relation that $\mathrm{E}[N(u,v)] \leq \mathrm{E}[N(u,u)]$. These results have been extended for estimating the expected fatigue damage rates for transformed Gaussian loads as well.[50,148,149]

5. Inverse Problems

The solution of inverse problems is much more computationally challenging than solving a forward problem. The difficulties associated with the identification of the system parameters from measurement data can be attributed to the noise that invariably exist in all measurements, erroneous calibration of the measurement sensors, incomplete measurement data, imprecise model for the system arising due to insufficient knowledge and lack of understanding about the physics associated with the system and incomplete knowledge about the parameters associated with the system.[150] These difficulties imply that inverse problems are ill-posed, often leading to situations with non-unique solutions or solutions that are physically unfeasible. A mathematically rigorous approach to the solution of inverse problems has been to employ the principles of Bayesian theories to estimate the unknown parameters in a system. The underlying principle of such frameworks is based on treating the model parameters to be identified as random variables or random fields, with assumed probability density functions. For a dynamical system expressed in the following general first order differential form

$$\dot{\mathbf{X}}(t) = \mathbf{f}(\mathbf{X}(t), \boldsymbol{\theta}(t), t), \tag{5.23}$$

$\boldsymbol{\theta}(t)$ represents the vector of system parameters to be identified. Here, $\mathbf{X}(t)$ is the measurable metric of the dynamical system, usually defined in terms of the state vector, the vector function $\mathbf{f}(\cdot)$ represents typically a nonlinear function available either explicitly or in implicit form and t is time. The pdf for $\boldsymbol{\theta}(t)$ at time instant t, is denoted by $p_{\boldsymbol{\theta}}(\boldsymbol{\theta}; t)$. Here, $p_{\boldsymbol{\theta}}(\boldsymbol{\theta}; t = t_0) \equiv p_{\boldsymbol{\theta}}(\boldsymbol{\theta})$ is referred to as the prior density function and takes into account information available about $\boldsymbol{\theta}$ at initial time $t = t_0$. If no information is available, a model for $p_{\boldsymbol{\theta}}(\boldsymbol{\theta})$ is assumed such that its support is defined within a practically feasible domain. The solution of the forward problem enables probabilistic characterization of the observable metric $\mathbf{X}$ at time t. Subsequently, using available measurement data $\mathbf{D}$, the posterior pdf for $\boldsymbol{\theta}$ can be expressed in terms of Bayes' theorem as

$$p(\boldsymbol{\theta}|\mathbf{D}) = \frac{p(\mathbf{D}|\boldsymbol{\theta}) p_{\boldsymbol{\theta}}(\boldsymbol{\theta})}{\int p(\mathbf{D}|\boldsymbol{\theta}) p_{\boldsymbol{\theta}}(\boldsymbol{\theta}) \, \mathrm{d}\boldsymbol{\theta}}. \tag{5.24}$$

Here, the conditional pdf $p(\mathbf{D}|\boldsymbol{\theta})$ is the normalized likelihood function. This procedure of updating the pdf of the system parameters can be carried out recursively for each time step for which measurements are available, with the *posteriori* pdf at time step t_k taken to be the prior density for time step t_{k+1}. This procedure of Bayesian updating ensures that the associated variability in the pdf decrease as more measurements are assimilated leading to fairly accurate estimates of the system parameters with quantifiable confidence bounds.

The development of the equations for estimating the *posteriori* pdf from available measurements and extracting relevant information from it constitute the essence of dynamic state estimation techniques. Typically, this involves the evaluation of multi-dimensional integrals whose dimensions equal the size of the vector $\boldsymbol{\theta}$. Closed form analytical solutions for these integrals are possible only for a special class of problems leading to the Kalman filter.[151] Methods which are variants of the Kalman filter, have been developed for parameter identification in problems in a more general setting.[152–158] However, these methods are usually iterative in nature, lacks universality in application and can be computationally expensive without a commensurate increase in robustness or accuracy. Alternative methods that rely on obtaining asymptotic approximations for the evaluation of these multi-dimensional integrals[159] or using numerical quadrature rules[160] have also been discussed.

However, the advent of inexpensive computing facilities has ensured the use of Monte Carlo simulations to approximate multi-dimensional integrals as a viable alternative.[161] This has led to the development of Monte Carlo based Bayesian algorithms,[162–171] commonly known as particle filters, for parameter identification from measurements. The primary advantages of particle filters lie in their general nature and wide applicability for problems even with high degrees of nonlinearity. These methods have been used for system identification in a wide variety of problems such as, climate modelling,[172] geophysics,[173,174] heat transfer,[175,176] and structural health monitoring.[53,177–184] Particle filters require the solution of the forward problem for a large number of realizations for $\boldsymbol{\theta}$, generated using Monte Carlo simulations and evaluating their likelihood when compared with the measurement data. This needs to be carried out for all the available measurements. The drawback of the particle filtering approaches lies in the computational costs involved in solving the forward problem a large number of times, corresponding to each measurement data set. This becomes computationally infeasible for complex problems where a solution of a single forward problem requires significant computational cost. This has led to studies being carried out which aim to reduce the associated computational cost. The central challenge lies in accelerating the solution of the forward problem for the set of sample realizations of $\boldsymbol{\theta}$. This can be achieved by developing surrogate models for the system such that the solution of the forward problem is simple. Response surface based methods[185–187] are simple to implement, however, they lack mathematical rigor and are not universally applicable. An alternative approach built on more rigorous mathematical foundations would be to use stochastic spectral methods to represent the structure response as a function of the unknown system parameters.

The use of PCE within Bayesian frameworks for parameter estimation in inverse problems is of recent vintage; see for example,.[52,188–195] These studies use PCE to analyse the propagation of uncertainty through a system where the unknown parameters have been modelled as random variables/random fields. Subsequently, estimates of the unknowns have been obtained from the posterior probability density functions which, in most studies, have been approximated by minimizing the covariance. Thus, these methods are based on the principles of Kalman filter and its variants[189,190,193–195] and involve linearizations or other forms of local approximations. In contrast, recent studies[52,53,188] use spectral methods in conjunction with particle filters and which are more generally applicable irrespective of the nonlinearity in the problem. Here, a surrogate model for the forward problem is built using PCE, and MCS are used for solving the multidimensional integrals.

It must be emphasized here that the limits to the dimensionality of the vector of parameters, $\boldsymbol{\theta}$, to be identified remains an open question. This is particularly relevant for system identification in the context of structural health monitoring where the dimension of $\boldsymbol{\theta}$ is usually large.[196] In the direct application of particle filtering for identification of system parameters, the dimension of $\boldsymbol{\theta}$ has been reported to be of the order of 50-60.[181,182] In replacing the forward problem using a PCE surrogate, each unknown parameter is expressed in terms of a PCE whose component projections are unknowns, thereby increasing the dimensionality of the problem. However, when the identification algorithm is transformed to the $\boldsymbol{\xi}$-space, the dimensionality of the problem is restricted to the dimension of $\boldsymbol{\xi}$. The questions on the limits of dimensionality of the unknown vector would therefore be identical to the problem of state estimation using the bootstrap particle filter.[162] More investigations need to be carried out in this context.

Often an UQ problem involves estimation of sensitivity of the response with respect to the random parameters, a topic not covered in this article. Sensitivity analysis is usually performed to identify the most dominant parameters. This information helps in reducing the stochastic dimensionality — thus the computational cost — of the problem. This can also help in economizing the quality control in a manufacturing setup. Sensitivity analysis can be classified into two types: local sensitivity, which is gradient-based, and global sensitivity, which is based on variance analysis such as Sobol' indices. This is a reasonably rich area, with newer developments still coming and being applied.

6. Parallelization

With the advent of affordable computational hardware, there is an increasing demand for harnessing parallel computing to perform UQ on large structures.[197] Here the word *large* qualifies the number of DOF, and not necessarily the physical size of the structure. For instance, a crack propagation problem in a physically small structure can easily involve a very large DOF. Load balancing and minimizing communication among processors are two major goals in a successfully implemented parallel solver. Because of the new computational hardware where interprocessor communication is also involved, development of parallel solvers often require entirely new approach compared to traditional solvers. One way to classify the computational problems is to look at the level coupling among different parts or subroutines. One extreme is the MCS or any of its variants. In this case all the realizations are independent of each other, thus can be run on different processors independently, thereby making the parallelization very simple. Similar independence holds for non-intrusive methods. On the other hand, intrusive solvers have a high level of coupling among the final system of algebraic equations. Developing parallel solvers for intrusive solvers is an active area of research.[198] Recent attempts to this end includes using domain decomposition.[199–203]

7. Current Challenges and Future Directions

Based on the review carried out, the following appears to be a set of current challenges:

(1) The developments in UQ have been mostly illustrated through simple numerical examples. In most cases, extending the algorithms and the methodologies to large scale engineering problems would become computationally intensive. Therefore, significant effort needs to be extended into development of reduced order modelling both in the number of state variables and in the associated stochastic dimensions.

(2) For problems with high degrees of nonlinearity, often Monte Carlo simulations, despite the high computational costs, offer the only hope for UQ. There is therefore a need to develop MCS based intelligent algorithms in conjunction with new parallelization techniques and algorithms for efficient UQ of stochastic nonlinear problems.
(3) Development of scalable parallel solvers for UQ is still in its early stage and a significant effort is needed in this area.

References

1. S. Ferson and L. Ginzburg, Different methods are needed to propagate ignorance and variability, *Reliab. Eng. Sys. Saf.* **54**, 133–144 (1996).
2. J. Helton and W. Oberkampf, Alternative representations of epistemic uncertainty, *Reliab. Eng. Syst. Saf.* **85**, 1–369 (2004).
3. JCSS, Probabilistic model code, Technical Report, Joint Committee on Structural Safety, http://www.jcss.ethz.ch/Home/RechteSeite.html (2001).
4. G. Stefanou and M. Papadrakakis, Stochastic finite element analysis of shells with combined random material and geometric properties, *Comput. Methods Appl. Mech. Engrg.* **193**(1), 139–160 (2004).
5. M. Ostoja-Starzewski, Random field models of heterogeneous materials, *Int. J. Solids Struct.* **35**(19), 2429–2455 (1998).
6. G. I. Schuëller, H. J. Pradlwarter and C. A. Schenk, Non-stationary response of large linear FE models under stochastic loading, *Comput. Struct.* **81**(8), 937–947 (2003).
7. A. M. Abbas and C. S. Manohar, Reliability-based critical earthquake load models. Part 1: linear structures, *J. Sound Vibration* **28**, 865–882 (2005).
8. K. R. Gurley, M. A. T. MA and A. Kareem, Analysis and simulation tools for wind engineering, *Probab. Eng. Mech.* **12**(1), 9–31 (1997).
9. A. Suryawanshi and D. Ghosh, Wind speed prediction using spatio-temporal covariance, *Natural Hazards* **75**(1435–1449) (2015).
10. T. Gneiting, Nonseparable, stationary covariance functions for space-time data, *J. Amer. Statist. Assoc.* **97**(458), 590–600 (2002).
11. C. S. Manohar and R. A. Ibrahim, Progress in structural dynamics with stochastic parameter variations; 1987–1998, *Appl. Mech. Rev.* **52**(5), 177–197 (1999).
12. C. S. Manohar and S. Gupta, Modeling and evaluation of structural reliability: current status and future directions, *Recent Advances in Structural Engineering, eds. K. S. Jagadish and R. N. Iyengar. University Press*, pp. 90–187 (2005).
13. A. Der Kiureghian and J.-B. Ke, The stochastic finite element method in structural reliability, *Probab. Eng. Mech.* **3**(2), 83–91 (1988).

14. B. Sudret and A. Der Kiureghian, Stochastic finite element methods and reliability: a state-of-the-art report, Technical Report Rep. No. UCB/SEMM-2000/08, University of California at Berkeley, USA (2000).
15. A. Haldar and S. Mahadevan, *Reliability Assessment Using Stochastic Finite Element Analysis.* Wiley (2000).
16. M. Kleiber and T. Hien, *The Stochastic Finite Element Method.* Wiley, Chichester (1992).
17. R. Ghanem and P. D. Spanos, *Stochastic Finite Elements: A Spectral Approach*, rev edn. Dover Publications (2003).
18. E. Vanmarcke and M. Grigoriu, Stochastic finite element analysis of simple beams, *J. Eng. Mech.* **109**, 1203–1214 (1983).
19. G. A. Fenton and E. H. Vanmarcke, Simulation of random fields via local average subdivision, *J. Eng. Mech.* **116**(8), 1733–1749 (1990).
20. A. Der Kiureghian and J. Ke, The stochastic finite element method in structural reliability, *Probab. Eng. Mech.* **3**(2), 83–91 (1988).
21. S. Mahadevan and A. Haldar, Practical random field discretization in stochastic finite element analysis, *Struct. Saf.* **9**(4), 283 – 304 (1991).
22. G. Deodatis and M. Shinozuka, The weighted integral method. II: Response variability and reliability, *J. Eng. Mech.* **117**(8), 1865–1877 (1991).
23. C.-C. Li and A. Der Kiureghian, Optimal discretization of random fields, *J. Eng. Mech.* **119**(6), 1136–1154 (1993).
24. S. Gupta and C. Manohar, Dynamic stiffness method for circular stochastic Timoshenko beams: response variability and reliability analyses, *J. Sound Vibration* **253**(5), 1051–1085 (2002).
25. H. Van Trees, *Detection, Estimation and Modulation Theory, Part I.* Wiley (2001).
26. B. A. Zeldin and P. D. Spanos, On random field discretization in stochastic finite elements, *ASME J. Appl. Mech.* **65**, 320–327 (1998).
27. S. P. Huang, S. T. Quek and K. K. Phoon, Convergence study of the truncated Karhunen–Loève expansion for simulation of stochastic processes, *Int. J. Numer. Methods Eng.* **52**(9), 1029–1043 (2001).
28. P. D. Spanos, M. Beer and J. Red-Horse, Karhunen-Loève expansion of stochastic processes with a modified exponential covariance kernel, *J. Eng. Mech.* **133**(7), 773–779 (2007).
29. R. Ghanem, The nonlinear Gaussian spectrum of log-normal stochastic processes and variables, *J. Appl. Mech.* **66**(4), 964–973 (1999).
30. K. K. Phoon, S. P. Huang and S. T. Quek, Simulation of second-order processes using Karhunen-Loève expansion, *Comput. Struct.* **80**(12), 1049–1060 (2002).
31. K. K. Phoon, H. W. Huang and S. T. Quek, Simulation of strongly non-gaussian processes using Karhunen–Loève expansion, *Probab. Eng. Mech.* **20**(2), 188–198 (2005).

32. K. E. Atkinson, *The Numerical Solution of Integral Equations of the Second Kind.* Cambridge University Press (1997).
33. G. Stefanou, The stochastic finite element method: Past, present and future, *Comput. Methods Appl. Mech Engrg.* **198**, 1031–1051 (2009).
34. D. C. Charmpis and M. Papadrakakis, Improving the computational efficiency in finite element analysis of shells with uncertain properties, *Comput. Methods Appl. Mech. Engrg.* **194**, 1447–1478 (2005).
35. P. Srikara and D. Ghosh, Faster computation of the Karhunen-Loève expansion using its domain independence property, *Comput. Methods Appl. Mech. Engrg.* **285**, 125–145 (2015).
36. D. L. Allaix and V. I. Carbone, Numerical discretization of stationary random processes, *Probab. Eng. Mech.* **25**, 332–347 (2010).
37. B. Khoromskij, A. Litvinenko, and H. Matthies, Application of hierarchical matrices for computing the Karhunen–Loève expansion, *Computing* **84**, 49–67 (2009).
38. C. Schwab and R. A. Todor, Karhunen–Loève approximation of random fields by generalized fast multipole methods, *J. Comput. Phys.* **217**(1), 100–122 (2006).
39. W. Betz, I. Papaioannou and D. Straub, Numerical methods for the discretization of random fields by means of the Karhunen–Loève expansion, *Comput. Methods Appl. Mech. Engrg.* **271**, 109–129 (2014).
40. R. Gutiérrez, J. Ruiz and M. Valderrama, On the numerical expansion of a second order stochastic process, *Appl. Stochastic Models Data Anal.* **8**(2), 67–77 (1992).
41. C. F. Li, Y. T. Feng, D. R. J. Owen, D. F. Li and I. M. Davis, A Fourier-Karhunen-Loève discretization scheme for stationary random material properties in SFEM, *Int. J. Numer. Methods Eng.* **73**, 1942–1965 (2008).
42. D. Xiu and G. E. Karniadakis, The Wiener-Askey polynomial chaos for stochastic differential equations, *SIAM J. Sci. Comput.* **24**(2), 619–64 (2002).
43. D. Xiu, D. Lucor, C.-H. Su and G. E. Karniadakis, Stochastic modeling of flow-structure interactions using generalized polynomial chaos, *J. Fluid Eng.* **124**(1), 51–59 (2002).
44. J. A. S. Witteveen, S. Sarkar, and H. Bijl, Modeling physical uncertainties in dynamic stall induced fluid–structure interaction of turbine blades using arbitrary polynomial chaos, *Comput. Struct.* **85**, 866–878 (2007).
45. D. Ghosh, R. Ghanem and J. Red-Horse, Analysis of eigenvalues and modal interaction of stochastic systems, *AIAA J.* **43**(10), 2196–2201 (2005).
46. R. Ghanem and D. Ghosh, Efficient characterization of the random eigenvalue problem in a polynomial chaos decomposition, *Int. J. Numer. Methods Eng.* **72**(4), 486–504 (2007).
47. D. Ghosh and R. Ghanem, An invariant subspace-based approach to the random eigenvalue problem of systems with clustered spectrum, *Int. J. Numer. Methods Eng..* **91**(4), 378–396 (2012).

48. D. Ghosh, Application of the random eigenvalue problem in forced response analysis of a linear stochastic structure, *Arch. Appl. Mech.* **83**, 1341–1357 (2013).
49. J. A. Witteveen, A. Loeven, S. Sarkar and H. Bijl, Probabilistic collocation for period-1 limit cycle oscillations, *J. Sound Vibration* **311**(1), 421–439 (2008).
50. S. Sarkar, S. Gupta and I. Rychlik, Wiener chaos expansions for estimating rainflow fatigue damage in vibrating structures with uncertain parameters, *Probab. Eng. Mech.* **26**, 387–398 (2010).
51. N. Ganesh and S. Gupta, Estimating the rain-flow fatigue damage in wind turbine blades using polynomial chaos, *J. Life Cycle Reliab. Saf. Eng.* **1**(4), 17–25 (2012).
52. Y. Marzouk, H. Najm and L. Rahn, Stochastic spectral methods for efficient Bayesian solution of inverse problems, *J. Comput. Phys.* **224**, 560–586 (2007).
53. P. Rangaraj, A. Chaudhuri and S. Gupta, The use of polynomial chaos for parameter identiifcation from measurements in nonlinear dynamical systems, *J. Appl. Math. Mech. (Z. Angew. Math. Mech)* **in press** (2015).
54. R. A. Todor and C. Schwab, Convergence rates for sparse chaos approximations of elliptic problems with stochastic coefficients, *IMA J. Numer. Anal.* **27**, 232–261 (2007).
55. A. Keese and H. G. Matthies, Numerical methods and Smolyak quadrature for nonlinear stochastic partial differential equations, Informatikbericht 2003-5, Institute of Scientific Computing, Department of Mathematics and Computer Science, Technische Universität Braunschweig, Brunswick (2003).
56. D. Ghosh and C. Farhat, Strain and stress computation in stochastic finite element methods, *Int. J. Numer. Methods Eng.* **74**(8), 1219–1239 (2008).
57. F. Nobile, R. Tempone, and C. Webster, An anisotropic stochastic collocation method for elliptic partial differential equations with random input data, Technical Report, MOX, Dipartimento di Matematica, Report No. 04/2007 (2007).
58. D. Ghosh and R. Ghanem, Stochastic convergence acceleration through basis enrichment of polynomial chaos expansions, *Int. J. Numer. Methods Eng.* **73**(2), 162–184 (2008).
59. X. Wan and G. E. Karniadakis, An adaptive multi-element generalized polynomial chaos method for stochastic differential equations, *J. Comput. Phys.* **209**, 617–642 (2005).
60. S. Sarkar and D. Ghosh, A hybrid method for stochastic response analysis of a vibrating structure, *Arch. Appl. Mech.* (Accepted, 2015, DOI 10.1007/s00419-015-1007-6).
61. D. Xiu, Fast numerical methods for stochastic computations: A review, *Communi. Comput. Phys.* **5**(2–4), 242–272 (2009).

62. J. B. Debusschere, H. N. Najm, P. P. Pébay, O. M. Knio, R. G. Ghanem and M. O. Le, Numerical challenges in the use of polynomial chaos representations for stochastic processes, *SIAM J. Sci. Comput.* **26**(2), 698–719 (2005).
63. A. Hasofer and N. Lind, Exact and invariant second moment code format, *J. Eng. Mech. (ASCE)* **100**, 111–121 (1974).
64. R. Rackwitz and B. Fiessler, Structural reliability under combined random load sequences, *Comput. Struct.* **9**, 489–494 (1978).
65. K. Breitung, Asymptotic approximations for multinormal integrals, *J. Eng Mech. (ASCE)* **110**(3), 357–366 (1984).
66. A. Naess, Bounding approximations to some quadratic limit states, *J. Eng. Mech. (ASCE)* **113**(10), 1474–1492 (1987).
67. K. Breitung, Asymptotic approximations for probability integrals, *Probab. Eng. Mech.* **4**(4), 187–190 (1989).
68. L. Tvedt, Distribution of quadratic forms in normal space: application to structural reliability, *J. Eng. Mech. (ASCE)* **116**(6), 1183–1197 (1990).
69. A. Der Kiureghian and M. Stefano, Efficient algorithm for second order reliability analysis, *J. Eng. Mech. (ASCE)* **117**(12), 2904–2923 (1991).
70. G. Cai and I. Elishakoff, Refined second-order reliability analysis, *Struct. Saf.* **14**, 267–276 (1994).
71. D. Polidori, J. Beck and C. Papadimitriou, New approximations for reliability integrals, *J. Eng. Mech. (ASCE).* **125**(4), 466–475 (1999).
72. Y. Zhao and T. Ono, New point estimates for probability moments, *J. Eng. Mech. (ASCE).* **126**(4), 433–436 (2000).
73. Y. Zhao and T. Ono, Moment methods for structural reliability, *Struct. Saf.* **23**, 47–75 (2001).
74. Y. Zhao and T. Ono, On the problems of fourth moment method, *Struct. Saf.* **26**, 343–347 (2004).
75. J. Marczyk, *Principles of Simulation Based Computer Aided Engineering.* FIM Publications, Barcelona (1999).
76. R. Rubinstein, *Simulation and the Monte Carlo Method.* Wiley Series in Probability and Mathematical Statistics, Wiley, New York (1981).
77. J. S. Liu, *Monte Carlo Strategies in Scientific Computing.* Springer Series in Statistics, Springer, New York (2008).
78. H. Kahn, Use of different Monte Carlo sampling techniques. In ed. H. Meyer, *Symposium on Monte Carlo Methods*, pp. 146–190. Wiley, New York (1956).
79. H. Kamal and B. Ayyub, Variance reduction techniques for simulation based structural reliability assessment of systems. In *8th ASCE Speciality Conference on Probabilistic Mechanics and Structural Reliability* (2000).
80. E. Gumbel, *Statistics of Extremes.* Columbia University Press (1958).
81. J. Galambos, *The Asymptotic Theory of Extreme Order Statistics.* Wiley, New York (1978).
82. E. Castillo, *Extreme Value Theory in Engineering.* Academic Press, Boston (1988).

83. S. Kotz and S. Nadarajah, *Extreme Value Distributions*. Imperial College Press, London (2000).
84. M. Pandey, P. Van Gelder and J. Vrijling, The estimation of extreme quantiles of wind velocity using l-moments in the peaks over threshold approach, *Struct. Saf.* **23**(2), 179–192 (2001).
85. R. Reiss and M. Thomas, *Statistical Analysis of Extreme Values*. Birkhäuser Verlag, Berlin (2001).
86. H. Cramer and M. Leadbetter, *Stationary and Related Stochastic Processes: Sample Function Properties and Their Applications*. Wiley, New York (1967).
87. S. O. Rice, A mathematical analysis of noise. In ed. N. Wax, *Selected Papers in Random Noise and Stochastic Processes*, pp. 133–294. Dover Publications (1956).
88. M. Marcus, Level crossings of a stochastic process with absolutely continuous sample paths, *Ann. probab.* **5**(1), 52–71 (1977).
89. T. Galtier, S. Gupta and I. Rychlik, Crossings of second-order response processes subjected to LMA loadings, *J. Probab. Statist.* **2010**, p. 752452 (2010).
90. Y. Wen, *Structural Load Modeling and Combination for Performance and Safety Evaluation*. Elsevier, Amsterdam (1990).
91. M. Grigoriu, Crossings of non-Gaussian translation processes, *J. Eng. Mech. (ASCE)* **110**(4), 610–620 (1984).
92. S. Winterstein, Non-normal responses and fatigue damage, *J. Eng. Mech. (ASCE)* **111**(10), 1291–1295 (1985).
93. S. Winterstien and O. Ness, *Hermite moment analysis of nonlinear random vibration*, In eds. W. Liu and T. Belytschko, *Computational Mechanics of Probabilistic and Reliability Analysis*, Chapter 21, pp. 452–478. Elme Press (1989).
94. M. Ochi and K. Ahn, Probability distribution applicable to non-Gaussian random processes, *Probab. Eng. Mech.* **9**(4), 255–264 (1994).
95. I. Rychlik, P. Johannesson and M. Leadbetter, Modeling and statistical analysis of ocean-wave data using transformed Gaussian processes, *Marine Struct.* **10**(1), 13–47 (1997).
96. U. Machado, Probability density functions for nonlinear random waves and responses, *Ocean Eng.* **30**(8), 1027–1050 (2003).
97. S. Kotz, T. Kozubowski and K. Podgorski, *The Laplace Distribution and Generalizations: A Revisit with Applications to Communications, Economics, Engineering and Finance*. Birkhäuser, Boston (2001).
98. J. Jith, S. Gupta and I. Rychlik, Crossing statistics of quadratic transformnations of lma processes, *Probab. Eng. Mech.* **33**, 9–17 (2013).
99. C. Gardiner, *Handbook of Stochastic Methods for Physics, Chemistry and the Natural Sciences*. Springer, Berlin (1983).
100. H. Risken, *The Fokker–Planck Equation: Methods of Solution and Applications*. Springer, New York (1996).

101. Y. Lin and C. Cai, *Probabilistic Structural Dynamics*. McGraw-Hill (2004).
102. J. Roberts and P. Spanos, Stochastic averaging: an approximate method of solving random vibration problems, *Int. J. Non-Linear Mech.* **2**, 111–134 (1986).
103. R. Booton, Nonlinear control systems with random inputs, *IRE Trans. Circuit Theory* **CT-11**, 9–19 (1954).
104. R. Iyengar and P. Dash, Study of the random vibration of nonlinear systems by the Gaussian closure technique, *J. Appl. Mech. ASME* **45**, 393–399 (1978).
105. A. Baratta and G. Zuccaro, Analysis of nonlinear oscillators under stochastic excitation by the Fokker–Planck–Kolmogorov equation, *Nonlinear Dynam.* **5**, 225–271 (1994).
106. R. Stratonovich, *Topics in the Theory of Random Noise*. Vol. 1, Gordon and Breach, New York (1963).
107. R. Khasminskii, A limit theorem for the solution of differential equations with random right-hand sides, *Theory Probab. Appl.* **11**, 390–405 (1966).
108. L. Lutes, Approximate technique for treating random vibration of hysteretic systems, *J. Acoust. Soc. Am.* **48**, 299–306 (1970).
109. K. Sobczyk and J. Trebicki, Maximum entropy principle in stochastic dynamics, *Probab. Eng. Mech.* **5**, 102–110 (1990).
110. S. Assaf and L. Zirkie, Approximate analysis of nonlinear stochastic systems, *Int. J. Control* **23**, 477–492 (1976).
111. S. Crandall, Non-Gaussian closure for random vibration of non-linear oscillators, *Int. J. Non-Linear Mech.* **15**, 303–313 (1980).
112. G. Er, An improved closure method for analysis of nonlinear stochastic systems, *Nonlinear Dynam.* **17**, 285–297 (1998).
113. R. Langley, A finite element method for the statistics of nonlinear random vibration, *J. Sound Vibration* **101**(1), 41–54 (1985).
114. B. Spencer and L. Bergman, On the numerical solution of the Fokker–Planck equations for nonlinear stochastic systems, *Nonlinear Dynam.* **4**, 357–372 (1993).
115. P. Kumar and S. Narayanan, Solution of Fokker–Planck equation by finite element and finite difference methods for nonlinear system, *Sadhana* **31**(4), 455–473 (2006).
116. P. Kumar, S. Narayanan and S. Gupta, Finite element solution of Fokker–Planck equation of nonlinear oscillators subjected to colored non-Gaussian noise, *Probab. Eng. Mech.* **38**, 143–155 (2014).
117. J. Sun and C. Hsu, The generalized cell mapping method in nonlinear random vibration based upon short-time Gaussian approximation, *J. Appl Mech ASME* **57**, 1018–1025 (1990).
118. J. Yu, G. Cai and Y. Lin, A new path integration procedure based on Gauss–Legendre scheme, *Int. J. Non-Linear Mech.* **32**, 759–768 (1997).
119. A. Naess and V. Moe, Efficient path integration method for nonlinear dynamics system, *Probab. Eng. Mech.* **15**, 221–231 (2000).

120. P. Kumar and S. Narayanan, Modified path integral solution of Fokker-Planck equation: response and bifurcation of nonlinear systems, *J. Comput Nonlinear Dynam, ASME* **5**, 0110004-1–0110004-12 (2010).
121. A. Masud and L. Bergman. Solution of Fokker-Planck equation by finite element and finite difference methods for nonlinear system. In *Proceedings of International Conference on Structural Safety and Reliability*, pp. 1911–1916 (2005).
122. A. Naess, Crossing rate statistics of quadratic transformations of Gaussian processes, *Probab. Eng. Mech.* **16**, 209–217 (2001).
123. A. Naess and H. Karlsen, Numerical calculation of the level crossing rate of second order stochastic Volterra systems, *Probab. Eng. Mech.* **19**, 155–160 (2004).
124. S. McWilliam, Joint statistics of combined first and second order random processes, *Probab. Eng. Mech.* **19**(1), 145–154 (2004).
125. S. Gupta and C. Manohar, Probability distribution of extremes of von Mises stress in randomly vibrating structures, *J. Vibrations Acoust. ASME* **127**, 547–555 (2005).
126. H. Madsen, Extreme value statistics for nonlinear load combination, *J. Eng. Mech. ASCE* **111**, 1121–1129 (1985).
127. M. Kac and A. Siegert, On the theory of noise radio receivers with square law detectors, *J. Appl. Phys.* **18**, 383–397 (1947).
128. A. Naess, A study of linear combination of load effects, *J. Sound Vibration* **129**(2), 83–98 (1989).
129. O. Ditlevsen, First outcrossing probability bounds, *J. Eng. Mech. ASCE* **110**(2), 282–292 (1984).
130. O. Hagen and L. Tvedt, Vector process out-crossing as parallel system sensitivity measure, *J. Eng. Mech. ASCE* **117**(10), 2201–2220 (1991).
131. R. Rackwitz, Computational techniques in stationary and non-stationary load combination, *J. Struct. Eng.* **25**(1), 1–20 (1998).
132. D. Veneziano, M. Grigoriu, and C. Cornell, Vector-process models for system reliability, *J. Eng. Mech. ASCE* **103**(EM3), 441–460 (1977).
133. B. Leira, Multivariate distributions of maxima and extremes for Gaussian vector processes, *Struct. Saf.* **14**, 247–265 (1994).
134. B. Leira, Extremes of Gaussian and non-Gaussian vector processes: a geometric approach, *Struct. Saf.* **25**, 401–422 (2003).
135. S. Gupta and C. Manohar, Development of multivariate extreme value distributions in random vibration applications, *J. Eng. Mech. ASCE* **131**(7), 712–720 (2005).
136. S. Gupta and P. van Gelder, Extreme value distributions for nonlinear transformations of vector Gaussian processes, *Probab. Eng. Mech.* **22**, 136–149 (2007).
137. K. Mohan, J. Jith and S. Gupta. Multivariate extreme value distributions for vector LMA processes. In eds. G. Deodatis, B. Ellingwood and D. Frangopol, *Safety, Reliability, Risk and Life-Cycle Performance of Structures and Infrastructures*, pp. 2841–2847. CRC Press, New York (2013).

138. K. Sobczyk, S. Wedrychowicz and B. Spencer, Dynamics of structural systems with spatial randomness, *Int. J. Solids Struct.* **33**(11), 1651–1669 (1996).
139. H. Madsen, S. Krenk and N. Lind, *Methods of Structural Safety.* Engelwood Cliffs, NJ, Prentice-Hall (1986).
140. R. Tovo, Cycle distribution and fatigue damage under broad-band random loading, *Int. J. Fatigue* **24**, 1137–1147 (2002).
141. S. Gupta and P. van Gelder, Probability distribution of peaks for nonlinear combination of vector Gaussian loads, *J. Vibrations Acoust., ASME* **130**, 031011:1–031011:12 (2008).
142. S. Gupta, N. Shabakhty and P. van Gelder, Fatigue damage in randomly vibrating jack-up platforms under non-Gaussian loads, *Appl. Ocean Res.* **28** (6), 407–419 (2006).
143. M. Matsuishi and T. Endo, Fatigue of metals subject to varying stress. In *Proceedings of Japan Society of Mechanical Engineers*, Jukvoka, Japan (1968).
144. I. Rychlik, A new definition of the rainflow cycle counting method, *Int. J. Fatigue* **9**, 119–121 (1987).
145. I. Rychlik, Rainflow cycles in Gaussian loads, *Fatigue Fracture Eng. Mater. Struct.* **5**, 57–72 (1992).
146. I. Rychlik, On the "narrow band" approximation for expected fatigue damage, *Probab. Eng. Mech.* **8**, 1–4 (1993).
147. I. Rychlik, Note on cycle counts in irregular loads, *Fatigue Fracture Eng. Materials Struct.* **16**, 377–390 (1993).
148. I. Rychlik and S. Gupta, Rainflow fatigue damage for transformed Gaussian loads, *Int. J. Fatigue* **29**, 406–420 (2007).
149. S. Gupta and I. Rychlik, Rain-flow fatigue damage due to nonlinear combination of vector Gaussian loads, *Probab. Eng. Mech.* **22**, 231–249 (2007).
150. P. Maybeck, *Stochastic Models, Estimation and Control.* Academic Press, UK (1979).
151. R. Kalman, A new approach to linear filtering and prediction problems, *J. Basic Eng. D ASME* **82**, 35–45 (1960).
152. L. Ljung, Asymptotic behavior of the extended Kalman filter as a parameter estimation for linear systems, *Proc. IEEE AC* **24**, 36–50 (1979).
153. M. Hoshiya and E. Saito, Structural identification by extended Kalman filter, *J. Eng. Mech. ASCE* **110**, 1757–1770 (1984).
154. R. Ghanem and M. Shinozuka, Structural system identifcation I: Theory, *J. Eng. Mech. ASCE* **121**(2), 255–264 (1995).
155. J. Li and J. Roberts, Stochastic structural system identification, *Comput. Mech.* **24**, 206–212 (1999).
156. D. Wang and A. Haldar, System identification with limited observations and without input, *J. Eng. Mech. ASCE* **123**(5), 504–511 (1997).
157. E. Wan and R. Merwe, The unscented Kalman filter for nonlinear estimation. In *Adaptive Systems for Signal Processing, Communications and Control Symposium IEEE* pp. 153–158 (2000).

158. R. Tipireddy, H. Nasrellah and C. Manohar, A Kalman filter based strategy for linear structural system identification based on multiple static and dynamic test data, *Probab. Eng. Mech.* **24**, 60–74 (2009).
159. M. Evans and T. Swartz, Methods for approximating integrals in statistics with special emphasis on Bayesian integration problems, *Statisti. Sci.* **10** (3), 254–272 (1995).
160. R. Cools and P. Dellaportas, The role of embedded integration rules in Bayesian statistics, *Statist. Comput.* **6**(245–260) (1996).
161. M. Evans and T. Swartz, *Approximating Integrals via Monte Carlo and Deterministic Methods.* Oxford University Press (2000).
162. N. Gordon, D. Salmond and A. Smith, Novel approach to nonlinear/non-Gaussian Bayesian state estimation, *IEE Proc. F* **140**(2), 107–113 (1993).
163. W. Gilks, S. Richardson and D. Spiegelhalter, *Markov Chain Monte Carlo in Practice.* Chapman & Hall (1996).
164. G. Kitagawa, Monte Carlo filter and smoother for non-Gaussian nonlinear state space models, *J. Comput. Graphics Statist.* **5**, 1–25 (1996).
165. H. Tanizaki, *Nonlinear Filters: Estimation and Applications.* Springer, Berlin (1996).
166. A. Doucet, N. de Freitas and N. Gordon, *Sequential Monte Carlo Methods in Practice.* Springer, New York (2001).
167. B. Ristic, S. Arulampallam and N. Gordon, *Beyond the Kalman Filter: Particle Filters for Tracking Applications.* Artech House, Berlin (2004).
168. N. Chopin, Central limit theorem for sequential Monte Carlo methods and its application to Bayesian inference, *Ann. Statist.* **32**(4), 2385–2411 (2004).
169. C. Manohar and D. Roy, Monte Carlo filters for identification of nonlinear systems, *Sadhana* **31**(4), 399–427 (2006).
170. S. Ghosh, C. Manohar, and D. Roy, Sequential importance sampling filters with a new proposal distribution for parameter identification of structural systems, *Proc. Royal Soc. London A* **464**, 25–47 (2008).
171. J. Park, N. Namachchivaya and H. Yeong, Particle filters in a multiscale environment:homogenized hybrid particle filter, *J. Appl. Mech.* **78**, 061001 (2011).
172. C. Jackson, M. Sen and P. Stoffa, An efficient stochastic Bayesian approach to optimal parameter and uncertainty estimation for climate model predictions, *J. Climate* **306**(25), 2828–2841 (2004).
173. W. Gouveia and J. Scales, Resolution of seismic waveform inversion: Bayes versus ocean, *Inverse Problems* **13**, 323–349 (1997).
174. A. Malinverno, Parsimonious Bayesian Markov chain Monte Carlo inversion in a nonlinear geophysical problem, *Geophys. J. Int.* **151**, 675–688 (2002).
175. J. Wang and N. Zabaras, Hierarchial Bayesian models for inverse problems in heat conduction, *Inverse Problems* **21**, 183–206 (2005).
176. J. Wang and N. Zabaras, Using Bayesian statistics in the estimation of heat source in radiation, *Int. J. Heat Mass Trans.* **48**, 15–29 (2005).

177. J. Ching, J. Beck and K. Porter, Bayesian state and parameter estimation of uncertain dynamical systems, *Probab. Eng. Mech.* **21**, 81–96 (2006).
178. V. Namdeo and C. Manohar, Nonlinear structural dynamical system identification using adaptive particle filters, *J. Sound Vibration* **306**, 524–563 (2007).
179. B. Radhika and C. Manohar, Reliability models for existing structures based on dynamic state estimation and data based asymptotic extreme value analysis, *Probab. Eng. Mech.* **25**, 393–405 (2010).
180. H. Nasrellah and C. Manohar, A particle filtering approach for structural system identification in vehicle-structure interaction problems, *J. Sound Vibration* **329**(9), 1289–1309 (2010).
181. H. Nasrellah and C. Manohar, Particle filters for structural system identification using multiple test and sensor data: a combined computational and experimental study, *Struct. Control Health Monitor.* **18**, 99–120 (2011).
182. H. Nasrellah and C. Manohar, Finite element method based Monte Carlo filters for structural system identification, *Probab. Eng. Mech.* **26**, 294–307 (2011).
183. B. Pokale and S. Gupta, Damage estimation in vibrating beams from time domain experimental measurements, *Arch. Appl Mech.* **84**, 1715–1737 (2014).
184. R. Rangaraj, B. Pokale, A. Banerjee and S. Gupta, Investigations on a partical filter algorithm methodology for crack identification in beams from vibration measurements, *Struct. Control Health Monitor.* **in press** (2015).
185. A. Khuri and J. Cornell, *Response Surfaces: Design and Analyses.* Marcel and Dekker, New York (1987).
186. S. Gupta and C. Manohar, An improved response surface method for the determination of failure probability and importance measures, *Struct. Saf.* **26**, 123–139 (2004).
187. S. Gupta and C. Manohar, Improved response surface method for time variant reliability analysis of nonlinear random structures under nonstationary excitations, *Nonlinear Dynam.* **36**, 267–280 (2004).
188. Y. Marzouk and H. Najm, Dimensionality reduction and polynomial chaos acceleration of Bayesian inference in inverse problems, *J. Computat. Phys.* **228**, 1862–1902 (2009).
189. E. Blanchard, A. Sandu and C. Sandu, Parameter estimation for mechanical systems via an explicit representation of uncertainty, *Engineering Computations.* **26**(5), 541–569 (2009).
190. E. Blanchard, A. Sandu and C. Sandu, Polynomial chaos based parameter estimation methods applied to a vehicle system, *J. Multi-body Dynam.* **224**, 59–81 (2010).
191. G. Saad and R. Ghanem, Robust structural health monitoring using a polynomial chaos based sequential data assimilation technique. In *III ECCOMAS Thematic Conference on Computational Methods in Structural Dynamics and Earthquake Engineering*, Corfu (2011).

192. J. Kolansky and C. Sandu, Generalized polynomial chaos based extended Kalman filter: improvement and expansion. In *9th Internatioanl Conference on Multibody Systems, Nonlinear Dynamics and Control*, Vol. 7A, p. V07AT10A019 (2013).
193. P. Dutta and R. Bhattacharya, Nonlinear estimation of hypersonic state trajectories in Bayesian framework with polynomial chaos, *J. Guidance Control Dynam.* **33**(6), 1765–1768 (2010).
194. B. Rosic, A. Litvinenko, O. Pajonk and H. Matthies, Sampling free linear Bayesian update of polynomial chaos representations, *J. Comput. Phys.* **231** (17), 5761–5787 (2012).
195. R. Madankan, P. Singla, T. Singh and P. Scott, Polynomial chaos based Bayesian approach for state and parameter estimations, *J. Guidance, Control Dynam.* **36**(4), 1058–1074 (2013).
196. E. Ntosios, C. Papadimitriou, P. Panetsos, G. Karaiskos, K. Perros and P. Perdikaris, Bridge health monitoring system based on vibration measurements, *Bull. Earthquake Eng.* **7**, 469–483 (2009).
197. E. Johnson, C. Proppe, B. Spencer Jr, L. Bergman, G. Szekely and G. Schueller, Parallel processing in computational stochastic dynamics, *Probab. Eng. Mech.* **18**, 37–60 (2003).
198. A. Keese and H. G. Matthies, Hierarchical parallelisation for the solution of stochastic finite element equations, *Comput. Struct.* **83**, 1033–1047 (2005).
199. A. Sarkar, N. Benabbou and R. Ghanem, Domain decomposition of stochastic PDEs: Theoretical formulations, *Int. J. Numer. Methods Eng.* **77**(5), 689–701 (2009).
200. D. Ghosh, P. Avery and C. Farhat, A FETI-preconditioned conjugate gradient method for large-scale stochastic finite element problems, *Int. J. Numer. Methods Eng.* **80**(6–7), 914–931 (2009).
201. W. Subber and A. Sarkar, Dual–primal domain decomposition method for uncertainty quantification, *Comput. Methods Appl. Mech. Eng.* **266**, 112–124 (2013).
202. W. Subber and A. Sarkar, A domain decomposition method of stochastic PDEs: An iterative solution techniques using a two-level scalable preconditioner, *J. Comput. Phys.* **257**, 298–317 (2014).
203. G. Stavroulakis, D. G. Giovanis, M. Papadrakakis and V. Papadopoulos, A new perspective on the solution of uncertainty quantification and reliability analysis of large-scale problems, *Comput. Methods Appl. Mech. Engrg.* **276**, 627–658 (2014).

Chapter 6

Uncertainty Quantification in Aeroelastic Problems

Sunetra Sarkar*, Harshini Devathi and W. Dheelibun Remigius

Department of Aerospace Engineering
Indian Institute of Technology Madras
Chennai 600036, India
**sunetra.sarkar@gmail.com, sunetra@iitm.ac.in*

This chapter reviews a couple of generic nonlinear aeroelastic systems and their response in the presence of parametric uncertainties. Polynomial chaos expansion has emerged as a popular uncertainty quantification tool in the areas of flow and fluid structure interactions as it can achieve higher computational efficiency compared to standard sampling techniques like Monte Carlo simulations, with good accuracy. However, there are challenges and the present chapter highlights some of the inadequacies of polynomial chaos expansion techniques. This is reviewed for a couple of generic aeroelastic systems considered in the chapter as they involve irregular response surfaces. Hence, polynomial chaos expansion may not be the most preferred tool for following the stochastic bifurcation behaviour in such systems. In this context, a recently proposed uncertainty quantification tool called the Probability Density Evolution method has been reviewed and outlined. This method does not depend on polynomial representation of uncertainties or interpolation based techniques but follows the response density function directly with time. As a result, sharp changes in the response in the random domain need not necessarily pose a problem. The technique is applied on the same aeroelastic examples successfully. Its performance and comparison with polynomial chaos technique are reviewed and commented using limited numerical results.

1. Introduction

Fluid structure interaction systems are capable of showing very rich dynamical behaviour in terms of various types of bifurcations. As our computational resources are expanding, we are able to go towards more and more high fidelity models and capture the dynamics of realistic engineering systems. It is interesting to understand the dynamical behaviour and resolve the paths to instabilities in fluid structure interaction systems in order to look for proper precursors to predict them for engineering systems. Nonlinear aeroelastic systems show different paths to instabilities through various bifurcations routes. However, tracking the corresponding routes could pose serious challenge to many existing uncertainty quantification tools as the system behaviour can become quite complex. Uncertainty quantification through rigorous stochastic modelling has opened up an interesting area of work for the present day aeroelasticians: understanding the stochastic bifurcation behaviour of a nonlinear aeroelastic system and doing that successfully with a reasonable amount of computational effort.

An engineering aeroelastic system would in general give a better performance closer to its instability boundaries, however, it could be risky without a complete understanding of its route to instability. With parametric uncertainties (modelled as stochastic processes), there could be significant qualitative changes in its dynamical behaviour. The uncertainty quantification tool to be employed should be robust enough to capture and characterize them. In the recent past, spectral uncertainty quantification tools like the Polynomial Chaos Expansion (PCE) method has been successfully used in various unsteady flows and fluid structure interaction studies. They are more efficient than conventional sampling based tools like the Monte Carlo Simulations (MCS), which has a slow convergence rate with the number of samples. The non-intrusive version of PCE, directly projects the solution on some chosen collocation points in the random domain and one would require a direct simulation only at those collocation points. The number of required collocation points are usually much less compared to MCS, and hence this is a computationally much cheaper alternative. As this approach does not modify the original governing equations, it enables the use of the deterministic solver directly. Hence the name non-intrusive.

The non-intrusive PCE approach has been applied to various aeroelastic applications successfully. For large dimensional random space, this would however demand a multidimensional collocation grid demanding considerable computational efforts. In order to avoid this penalty (commonly known as 'curse of dimensionality'), sparse grid based approaches have been proposed. However, to resolve complex dynamical behaviour or response near a bifurcation or instability onset, the required number of data points could still be unrealistically high. This has triggered the growth of various adaptive tools in the PCE family to cater to specific dynamical scenarios.

However, they mostly cater to specific bifurcation patterns and may not be very efficient when applied to a different dynamical behaviour. For the want of a universal tool to follow the response throughout the stochastic bifurcation path, it would be better to opt for an approach which would track the system in terms of a tangible stochastic output directly, say the probability density or distribution functions. Such quantities enable us to quantify stochastic bifurcation behaviour directly. The present chapter reviews one such uncertainty tool, called the Probability Density Evolution Method (PDEM) which deals directly with the probability density function and follows its evolution with time. Only recently this tool has been applied successfully to aeroelastic examples which we review in this chapter and also compare its performance with PCE based approaches.

2. Uncertainty Quantification in Aeroelastic Systems Using Polynomial Chaos Expansion

It is increasingly being acknowledged in the aeroelastic community that aeroelastic analysis should include the effects of system parametric uncertainties. This can mark a radical change in the present design concepts with higher rated performance. Dynamical systems are known to be sensitive to physical uncertainties. Hence, quantifying the effect of uncertainty propagation on the stability boundary is crucial. The importance of modelling these uncertainties with stochastic tools is that they quantify the effect of uncertainties in a probabilistic sense and gives the response statistics in a systematic manner. Based on the resulting detailed probabilistic information, decisions can be made on the design optimization process.

There are various physical agencies with random characteristics that act on engineering fluid structure interaction systems. These can come from earthquakes, wind load, wave load, pressure variations in impeller cavities, and so on. They need to be described as stochastic processes. Moreover, in some cases, the system properties themselves are uncertain, and must be considered as stochastic processes, which give rise to a problem of multiplicative or parametric excitation. As a result, the differential equations that govern the response become stochastic differential equations.

In the recent past, Polynomial Chaos Expansion (PCE)[1,2] approach has been used successfully in various fluid structure interaction systems.[3–8] It uses a spectral representation of a stochastic process in terms of orthogonal polynomials, in which the stochastic quantity of interest is represented spectrally by employing orthogonal polynomials from the Askey scheme[9] as a basis in the random space. By incorporating this series expansion into the stochastic differential equation followed by an orthogonal projection onto the various basis functions, one can elicit differential equations without any random component which can be used to obtain the deterministic coefficients. The original homogeneous PCE was based on Hermite polynomials from the Askey family.[1,10] It gives optimal exponential convergence for Gaussian inputs.[11] A standard Galerkin projection is applied along the random dimensions to obtain a weak form of the equations.[12–14]

Galerkin polynomial chaos expansion — which is an intrusive approach — modifies the governing equations to a coupled form in terms of the chaos coefficients. These coupled equations are usually more complex and deriving them might prove a tedious task for some uncertain parameter choices. In order to avoid this, several uncoupled alternatives have been proposed. These are collectively called as non-intrusive approaches. Here, a deterministic solver is used repeatedly as in a Monte Carlo simulation at certain representative collocation points in the parametric range.[15–20] Non-intrusive spectral projection (NISP) based PCE algorithm has been applied extensively in aeroelastic systems[21–23] as well. When multiple uncertain parameters are involved, collocation grids are constructed using tensor products of one-dimensional grids.[24] Thus, the number of collocation points and therefore the number of required deterministic solves increase rapidly.

For spanning the random domain in terms of a number of independent random variables in spectral techniques, a random process needs to be discretized. The Karhunen-Loève expansion method[2,25] provides an efficient means to transform a random process into a set of independent random variables. In the PCE framework, these variables span the various dimensions of the input random space on which the orthogonal basis functions and response surfaces are constructed. If $\boldsymbol{\xi}$ denotes a set of random variables $\{\xi_1, \xi_2, \ldots, \xi_d\}$ used in the Karhunen–Loève expansion (will be discussed later in the chapter), then, using the PCE method, some system response variable $\alpha(\tau, \boldsymbol{\xi})$ can be written as,[26]

$$\alpha(\tau, \boldsymbol{\xi}) = \sum_{j=1}^{n} \widehat{\alpha}_j(\tau)\Phi_j(\boldsymbol{\xi}), \tag{6.1}$$

where polynomials Φ_j are the basis functions; $\widehat{\alpha}_j$ are the PCE coefficients and τ is the time parameter. The choice of the basis function depends on the random variables $\boldsymbol{\xi}$. The basis polynomials are chosen so as to be orthogonal with respect to $p_{\boldsymbol{\xi}}$, which is the joint probability density function of all the input random variables. It may be noted that the speed of convergence also depends on the choice of basis.[27]

The truncated series is considered up to a chosen order n which is called the order of chaos expansion; n should be large enough to ensure that the series gives converged solutions. For any non-Gaussian random variables, one can use orthogonal polynomials from the generalized Askey scheme for some standard non-Gaussian input distributions[12,13] or employ a Gram-Schmidt orthogonalization to generate an orthogonal family of polynomials.[28,29] In the non-intrusive spectral projection (NISP) approach, the deterministic coefficients are obtained by projecting the chaos expansion given by Eq. (6.1) onto the various basis functions. The basis functions (Hermite polynomials for Gaussian random variables) are statistically orthogonal, that is, they satisfy $\langle \Phi_i, \Phi_j \rangle = 0$ for $i \neq j$, hence the expansion coefficients can be directly evaluated as follows,[22,23,30,31]

$$\widehat{\alpha}_j(\tau) = \frac{\langle \alpha(\tau, \boldsymbol{\xi}), \Phi_j \rangle}{\langle {\Phi_j}^2 \rangle}. \tag{6.2}$$

The key step in projecting $\alpha(t, \boldsymbol{\xi})$ along the polynomial chaos basis is the evaluation of $\langle \alpha, \Phi_j \rangle$. For example, for a single Gaussian random variable case (ξ_1),

$$\langle \alpha(\tau, \xi_1), \Phi_j(\xi_1) \rangle = \int_{-\infty}^{+\infty} \alpha(\tau, \xi_1), \Phi_j(\xi_1) \mathcal{W}(\xi_1) d\xi_1, \tag{6.3}$$

where the weighting function $\mathcal{W}(\xi_1)$ is the Gaussian probability density function. For zero mean and unit variance case, this is given by the following,

$$\mathcal{W}(\xi_1) = \frac{1}{\sqrt{2\pi}} e^{-\frac{1}{2}{\xi_1}^2}. \tag{6.4}$$

The evaluation of $\langle \alpha(\tau, \xi_1), \Phi_j \rangle$ can be optimally done by using a Gauss-Hermite numerical integration scheme.

$$\begin{aligned} \langle \alpha(\tau, \xi_1), \Phi_j(\xi_1) \rangle &= \int_{-\infty}^{\infty} \alpha(\tau, \xi_1) \Phi_j(\xi_1) \mathcal{W}(\xi_1) d\xi_1 \\ &= \sum_{k=1}^{n} \{\alpha(\tau, \xi_1) \Phi_j(\xi_1)\}_k w_k, \end{aligned} \tag{6.5}$$

where $\{\alpha(\tau, \xi_1)\Phi_i(\xi_1)\}_k$ are evaluated at the various collocation points $\{\xi_1\}_k$ and w_k are the corresponding weights.[32,33] The collocation points are the roots of the nth order Hermite polynomial. At these points, the corresponding samples of the uncertain parameter is used to run a pseudo-MCS and the response realizations $\alpha(\tau, \xi_1)$ are then used to estimate the deterministic coefficients. For non-Gaussian random variables the associated integrals are evaluated by carefully chosen quadrature points and weights associated with the suitable chaos polynomials.[28,29,32,34] For higher dimensional random spaces such as those spanned by $\boldsymbol{\xi}$, the evaluation of the integrals which are multi-dimensional would require a multidimensional grid of collocation points which can be evaluated by taking the tensor products.[35] It should be noted that for irregularities or discontinuities in the response in the random domain, the efficiency of the quadrature schemes would suffer.

2.1. *Latest developments in sparse grid algorithms*

Certain aeroelastic systems like stall induced oscillations of wings or rotors can display strong nonlinear behaviour. Parametric uncertainties in such systems could translate into steep solution fronts in the random domain spanned by the random variables, especially in the vicinity of the deterministic bifurcation points. One would require a very high resolution while discretizing the random space in order to maintain accuracy of the stochastic response, however, at the expense of computational efforts. To counter this, the use of higher resolution only near the sharp solution changes has been suggested as a trade-off. To this end, various multi-element adaptive schemes have been developed to investigate the aeroelastic behaviour under the influence of parametric randomness.[36–38] In the multi-element adaptive schemes, to start with, the random space is divided into a minimum number of elements. Each element is associated with an error parameter whose value if exceeds a user-defined threshold limit, the corresponding element is split further. In a classical flutter system, Chassaing *et al.*[38] have divided the tensor grid under study into a number of sub-domains and the contribution of each of the elements to the global variance is used as an indicator for splitting the elements further. Witteveen *et al.*[36] have employed multi-element adaptive mesh refinement based on Newton-Cotes quadrature in simplex elements and the eigenvalues of the Hessian matrix of the PCE approximation of the solution has been used to evaluate the refinement measure. Witteveen and Bijl[37] have applied an adaptive meshing scheme to a PCE technique based on constant-phase interpolation which has been successfully applied in a classical aeroelastic problem.[23] In general, such adaptive tools are based on tensor grid based discretization which works quite efficiently for relatively lower dimensional random domains. For larger random domains, the required number of realizations may still be quite high as the number of realizations required increases exponentially with the dimension.

An answer to this problem lies with the sparse grid based collocation algorithms. The algorithm was originally developed by a Russian mathematician Sergey A. Smolyak.[39] The Smolyak algorithm uses the combination of a number of lower order quadrature rules to construct a higher order interpolant.[35,39,40] This ensures a controlled growth of the collocation points with the number of dimension. With a minimal loss in accuracy,[39,41] the number of collocation points for the sparse grids is drastically reduced to $\mathcal{O}(N(\log N)^{d-1})$ from $\mathcal{O}(N^d)$, which is the number of collocation points required for full tensor grids. Here, d represents the number of random dimensions and N represents the expansion order along each of them. Adaptive mesh refinement schemes on sparse grids were also developed by researchers.[42,43] The schemes employed by Ma and Zabaras[42] and Bungartz and Dirnstorfer[43] are point-wise techniques and are better suited for sparse grids. The difference between the true value of the function and the value obtained using a lower level quadrature, is used as indicators for adding the neighbouring points in a higher level quadrature scheme. This provides an efficient error estimations as it reduces to zero for smooth functions while ensuring sufficient resolutions for regions spanned by sharp solution fronts.[42] Due to the local nature of the schemes, the basis functions also should possess a local character and thus the global Lagrange, Hermite, Legendre and other polynomials with global character are unsuitable. Piecewise continuous, linear hat functions have also been tested successfully as basis functions[42] which can be applied only in the case of equidistant nodes. For non-equidistant nodes, the hierarchical Lagrange basis has been employed successfully.[43] The hierarchical Lagrange basis functions are evaluated similar to that of Lagrange polynomials, except that the support nodes are the set of ancestors from which the current node has evolved. As the evaluation of hierarchical Lagrange basis requires a history of all the ancestral nodes from which the current node has evolved, it can become slightly involved. Recently, an improved sparse grid algorithm, with the use of a modified Lagrange interpolating function has been applied successfully by one of the present authors in a stall induced aeroelastic system with multi-parametric uncertainties.[44,45] Only the neighbouring points were chosen as the support nodes making it suitable for point-wise adaptive meshing schemes. However, it was acknowledged that if there are sharp changes in the response surface behaviour, due to the underlying dynamics or bifurcations, sparse algorithms could also become inefficient requiring a large number of quadrature locations.[45]

3. Some Outstanding Issues with PCE

The application of PCE and its family of adaptive variants in aeroelastic problems is quite recent. However, stochastic modelling of aeroelastic systems has been actively pursued even earlier, mostly using conventional MCS techniques.[46,47] The method of stochastic averaging was also used to study the problem of nonlinear panel flutter and flexible helicopter rotor blade flapping by Namachchivaya and his co-workers.[48,49] Commonly, stochastic bifurcation is discussed in terms of the qualitative changes in the response probability density functions (PDF) as a bifurcation parameter is varied.

Till date, the most significant contribution in the area of stochastic bifurcations[50] applied to aeroelastic systems have been made by Poirel and Price[51–54] in a series of studies. Stochastic bifurcation in a nonlinear aeroelastic system has been studied for a random input wind speed. They have used the bifurcation of the response probability density function to follow the stochastic bifurcation. This has been termed as the phenomenological bifurcation (P-bifurcation) and has been compared with the dynamics bifurcation (D-bifurcation) in terms of the largest Lyapunov exponent. The onset of P-bifurcation could be ahead of that in a deterministic system. This earlier onset of the aeroelastic instability has been observed with system parametric randomness as well.[22,29,55] Other quantitative measures based on time series tools[56,57] have also been developed in the literature and have been implemented quite recently to predict qualitative changes in aeroelastic dynamics successfully.[58] Predictions from these tools could differ from the P-bifurcation onsets highlighting the need for more in depth understanding of the stochastic bifurcation behaviour.[59] The strengths and limitations of existing popular UQ tools in tracking the stochastic aeroelastic systems especially near the bifurcation points are therefore taken up here.

3.1. *Aeroelastic system with subcritical nonlinearity*

3.1.1. *Governing equation and PCE performance*

Systems showing subcritical variants of bifurcations are always susceptible to risk as they experience sudden sharp changes in their response. An example was presented in Patel *et al.*[60] for an aeroelastic system with uncertainties. A schematic is shown in Fig. 6.1(a) for a classical pitch-plunge aeroelastic system. Such dynamical systems with a softening spring (negative cubic nonlinearity) generically involve an unstable LCO and a bigger stable LCO in the phase space at the subcritical regime. The structural part in the equation has a softening spring which is responsible for the subcritical behaviour here. The governing equations of the system are available in the literature which are reproduced below using the standard notations:[61,62]

$$\begin{aligned} \bar{h}'' + x_\alpha \alpha'' + 2\zeta_{\bar{h}} \frac{\bar{\omega}}{U} \bar{h}' + \left(\frac{\bar{\omega}}{U}\right)^2 \bar{h} &= -\frac{1}{\pi\mu} C_L(\tau), \\ \frac{x_\alpha}{r_\alpha^2} \bar{h}'' + \alpha'' + 2\frac{\zeta_\alpha}{U} \alpha' + \frac{1}{U^2}(\alpha + \beta\alpha^3) &= \frac{2}{\pi\mu r_\alpha^2} C_M(\tau), \end{aligned} \tag{6.6}$$

where U is the nondimensional form of wind velocity, nondimensionalized using standard techniques.[63] The parameter values of this system has been

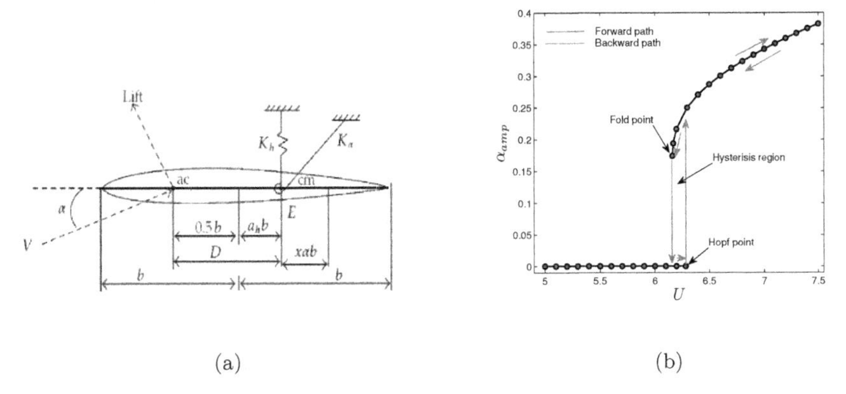

Fig. 6.1. (a) Schematic of a classical pitch-plunge aeroelastic system, (b) deterministic bifurcation diagram (subcritical) with wind velocity U as the bifurcation parameter.

taken from Lee *et al.*[61] and Alighanbari.[64] A nondimensional fifth order stiffness of 60 is also implemented here in pitch (not shown in the equation) to ensure a finite amplitude limit cycle oscillation. The deterministic bifurcation diagram is given in Fig. 6.1(b) which shows a subcritical Hopf bifurcation behaviour.

The subcritical dynamical behaviour of the aeroelastic system can be exactly modelled using smaller order dynamical systems. We consider one that is presented below to investigate the shortfall of PCE. The equation can be written as,[65,66]

$$\dot{R} = \mu R + \alpha R^3 - \beta R^5, \tag{6.7}$$

where R is the dynamical system variable and μ, α, and β are parameters that determine the behaviour of the system. Here R may be seen as the amplitude of evolution of an aeroelastic system. For the system to exhibit oscillatory behaviour an equation for the evolution of phase can be defined as:

$$\dot{\theta} = \omega + \kappa R^2, \tag{6.8}$$

where ω and κ are arbitrary constants. Such a subcritical nonlinear system would show sharp changes or 'jumpiness' in its response surface behaviour. 'Jumpy' response surfaces are difficult to model using PCE tools and one would require infinitely large order of expansion to capture the discontinuities accurately,[66,67] which is of course unfeasible. The system exhibits subcritical Hopf bifurcation as is known from the dynamical systems theory.[65] Thus, there exists a range of parameter μ for which two solutions are possible depending on the initial conditions — the fixed point solution or limit cycle oscillations. This is termed as the subcritical range of parameters in which we would present the performance of PCE. Variations in limit cycle amplitudes and chances of system failure would be of interest to aeroelasticians when uncertainty in one or more of the system parameters is considered. In the subcritical regime, the response depends critically on initial conditions as illustrated in Figs. 6.2(a) and 6.2(b). A small

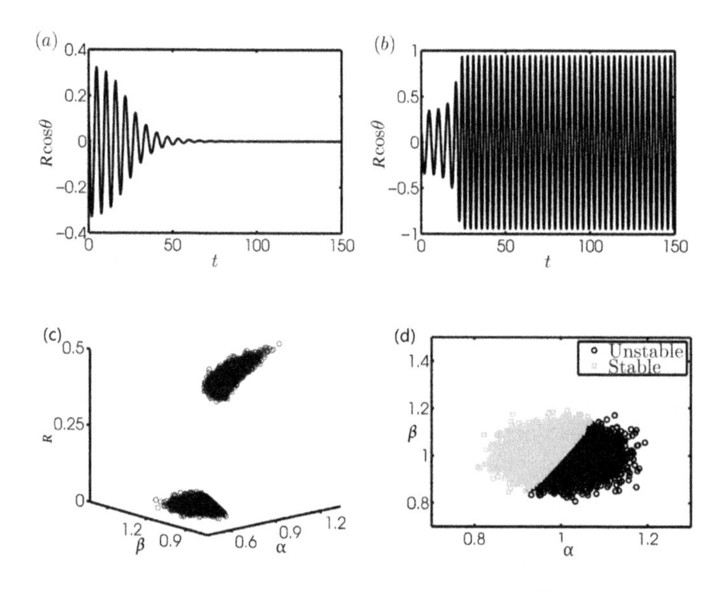

Fig. 6.2. (a)Time history of damped oscillation, (b) Time history of limit cycle oscillation (LCO), (c) 3D response surface for various random value of parameters α and β for $\mu = -0.25$, $R(0) = 0.71$, $\bar{\alpha} = 1.0$, $\sigma_a = 0.05$, $\bar{\beta} = 1.0$, $\sigma_b = 0.05$, $\kappa = 1.0$ (d) projection of 3D response surface on $\alpha - \beta$ plane.[66]

change in the initial perturbation can cause a sudden qualitative change in the behaviour of the system response eventually leading to limit cycle oscillations.

Figures 6.2(c) and 6.2(d) shown the nature of the 3D response surface and its projection at $\mu = -0.25$ for Gaussian random parameters α and β as shown. The plot was obtained using a standard MCS of 20000 samples for the same set of initial condition mentioned in the figure caption. The dark regions in the projected plot correspond to the region of instability (LCO) and light coloured regions indicate the regions of stability. The boundary represents the region of discontinuity where the nature of the response changes. The mean values of parameters α and β were chosen (mentioned along with Fig. 6.2) so as to obtain a limit cycle (instability) response. Without parametric uncertainty, the stability diagram would consist of just one unstable point with the coordinates corresponding to the mean values of the parameters. A system exhibiting subcritical bifurcations show discontinuities in their response surface behaviour and to capture the discontinuity, one would require an infinite polynomial order for PCE. Hence, methods involving polynomial expansions practically fail in such situations as is shown in Fig. 6.3 for a single random variable α. In Fig. 6.3(a), MCS results show a sudden jump which comes due to the subcritical behaviour.[66,68] The results of PCE are shown in the Fig. 6.3(b). A non-intrusive Polynomial Chaos Expansion (employing NISP) has been done for $\mu = -0.25$ which lies at the boundary of the subcritical zone. The uncertainty in α with a normal distribution is considered with $\bar{\alpha} = 1$, $\sigma_\alpha = 0.05$. Expansions in the response variable R were taken up to 9th order (10 terms in the expansion). A Hermite quadrature was employed (in projection based non-intrusive PCE) to calculate the inner products. As can be seen in Figs. 6.3(b) and 6.3(d), PCE is unable to capture the discontinuity and it could only be possible, if at all, with an infinitely large order of expansion. The response surface and the CDF (CDF brings out the differences better) look entirely different from those obtained using standard MCS.

The performance of PCE for the aeroelastic system given in Eq. (6.6) using the NISP scheme is shown in Fig. 6.4. The cubic stiffness parameter β is considered to be random with a uniform distribution. The actual response surface and that modelled by PCE of increasing order shown in the figure demonstrates the inadequacy of the polynomial based scheme in capturing the sharp change in the response surface.

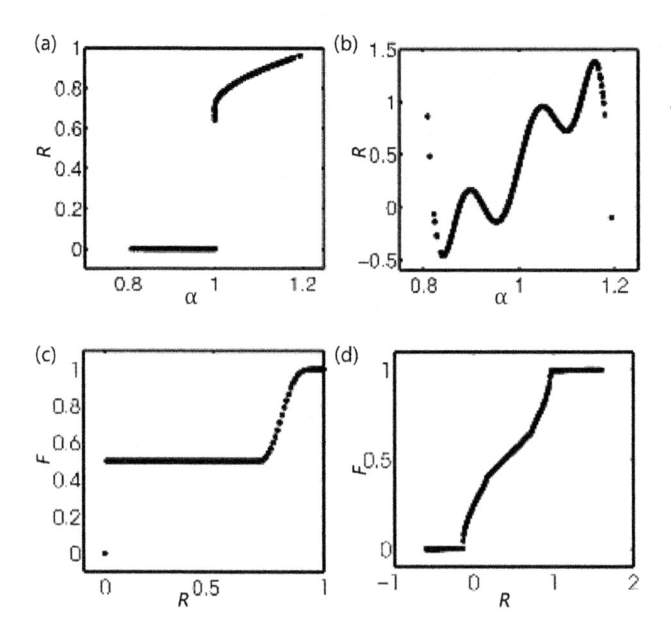

Fig. 6.3. A plot of the response surface of R for uncertainty in α with $\mu = -0.25$, $R(0) = 0.71$, $\bar{\alpha} = 1.0$, $\sigma_\alpha = 0.05$, $\beta = 1.0$ using (a) MCS (20000 deterministic solves), (b) non-intrusive PCE (order 9). The corresponding CDFs obtained are given in (c) and (d).[66]

3.2. *Aeroelastic system under external stochastic loading*

In addition to parametric uncertainties, an aeroelastic system is also subject to random, external velocity field fluctuations which could arise from a variety of sources such as atmospheric flow-field fluctuations, interactions from nearby structural components etc. These random fluctuations with a multiplicative coupling could bring a shift in the bifurcation point for nonlinear dynamical systems[59,69] or can show a very different qualitative behaviour. In addition, aeroelastic systems subjected to such external fluctuations like random wind-field can also suffer from irregular response surface behaviour. In fact, depending on the complexity of the wind, the occurrence of the irregularities can be quite frequent rendering PCE or recently proposed interpolation based schemes[68,70] ineffective. In this context, we review the performance of a stall induced aeroelastic system here. A stall aeroelastic system faces a complex aerodynamic behaviour due to the highly separated and unsteady flow-field near the wing or the airfoil. The corresponding stall models available in the literature which addresses

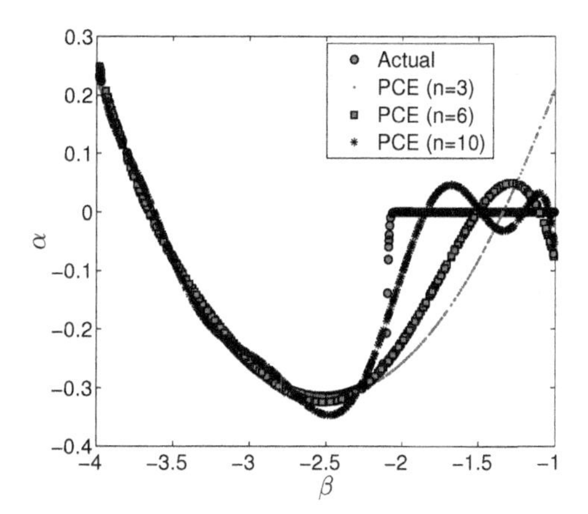

Fig. 6.4. Inadequacy of PCE (NISP) in the aeroelastic system with subcritical nonlinearity: pitch response α vs. random β with $\bar{\beta} = -3$ with a uniform distribution; $U = 6.2$; rest of the parameters are as listed by Lee *et al.*[61]

the complex interaction behaviour between the flow-field vortices through a set of differential equations[71–73] would have nonlinearities in complex forms. The corresponding stochastic response surface can manifest multiple sharp changes or irregularities.[45] To demonstrate this and also the performance of PCE, a model wind gust is taken which will be input to a stall induced aeroelastic system.

3.2.1. *A model wind gust for input*

A model wind gust ($\bar{V} + v$, where v is the fluctuating part) is modelled as a random process; an Ornstein-Uhlenbeck spectrum is considered here which is a stationary Gaussian random process and generally used in modelling the motion of a Brownian particle.[74] The autocorrelation function and the power spectral density of this stochastic process are respectively given as below.[74]

$$R_{vv,OU}(s) = \exp\big(-\eta|s|\big), \quad S_{vv,OU}(\omega) = \frac{2}{\pi}\frac{\eta}{\eta^2+\omega^2}, \tag{6.9}$$

where s is defined as $|s| = |t_i - t_j|$ and t_i and t_j are two different time instants given in seconds. $1/\eta$ is the autocorrelation length and ω is the angular frequency. Most of the wind data in literature are available as power spectral densities with no closed form analytical expressions for the corresponding autocorrelation functions (examples include, Harris spectrum,

Davenport spectrum, Von Karman spectrum etc.). To avoid a numerical computation of the autocorrelation matrix which could be computationally challenging, a modification in the above spectrum is done to be consistent with the real-life Harris wind spectrum:[75]

$$S_{vv,HAR}(\omega) = \frac{1}{\pi} \frac{2\kappa L u \eta}{\left(\eta^2 + \omega^2\right)^a} \frac{\eta^{2a-1}}{2^a}, \tag{6.10}$$

where, parameter $a = 5/6$, κ is the roughness coefficient of the terrain, L is the characteristic length, ω is the angular frequency, $\eta = (2\sqrt{2}\pi u)/L$ and u is the mean gust velocity. The above form can be simplified to be equivalent to an Ornstein-Uhlenbeck process of Eq. (6.9). The power spectral density and the autocorrelation function (analytically obtained) of this modified Ornstein-Uhlenbeck form are given below.

$$S_{vv,OUH}(\omega) = \frac{1}{\pi} \frac{2\kappa L u \eta^2}{\left(\eta^2 + \omega^2\right)} \quad R_{vv,OUH}(s) = \kappa L u \eta \exp\big(- \eta |s|\big). \tag{6.11}$$

These steps are not mandatory, however they simplify the numerical efforts considerably, especially for computationally intensive aeroelastic simulations.[76]

3.2.2. *Discretization of the input wind*

The discretization of the input wind in terms of a finite number of input random variables is necessary to perform a PCE to propagate the effect of the input. This is done using a Karhunen-Loève (KL) expansion technique.[2] The analytically obtained autocorrelation function (such as Eq. 6.11) is quite ideal to be used in the KL form. The Karhunen-Loève expansion used for discretizing a stochastic process can be employed only if the stochastic process is mean square continuous.[25] A stochastic process can be shown to be mean square continuous if its autocorrelation function is continuous[12,25] as is the present case. In the KL form, a Gaussian random process can be expanded in a series of standard Gaussian random variables and a set of orthonormal basis obtained from an eigenvalue decomposition of the corresponding covariance matrix and thus it may be thought of as a mapping from the random process domain to the random variable domain. The KL expansion (in the truncated form using d dominant eigenvalues) for the input stochastic process v with mean u can be expressed as,[2]

$$v(t) = u + \sum_{i=1}^{d} \sqrt{\lambda_i} \psi_i(t) \xi_i, \tag{6.12}$$

where, λ_i's and ψ_i's are the eigenvalues and the eigenvectors of the autocorrelation matrix, ξ_i's are the random variables. The Karhunen-Loève series is required to be truncated at an optimum number of terms in order to avoid an explosive growth of the collocation points associated with higher random dimensions. This can be accomplished by retaining only the significant eigenmodes of the autocorrelation matrix. The order of truncation study for the series of eigenvalues, $\lambda_1 > \lambda_2 > \cdots$ is carried out following the ratio of the magnitude of the dth eigenvalue to the total sum of all the eigenvalues. The truncation at a chosen number of d eigenvalues would result in d independent random variables in the subsequent PCE formulation.

3.2.3. *Governing equations and PCE performance*

The governing equation for stall aeroelastic system is shown below in the standard nondimensional form[77] and the symbols have their standard meanings.[63]

$$\alpha'' + \frac{2\zeta}{U}\alpha' + \frac{1}{U^2}\left(\alpha + \beta\alpha^3\right) = \frac{2C_m}{\pi\lambda r_\alpha^2}. \tag{6.13}$$

The wind here is modelled as a Gaussian random process as discussed in the earlier section. The aeroelastic system is shown schematically in Fig. 6.5 (a). The stall model used for the aerodynamic moment at the right hand side has been based on the Beddoes-Leishman calculations.[73] This has

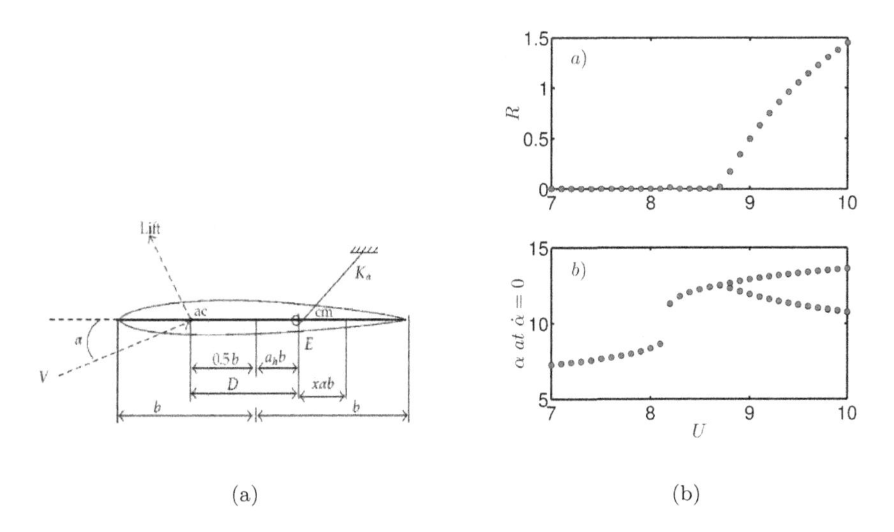

Fig. 6.5. (a) Schematic of a stall aeroelastic system, (b) deterministic bifurcation diagram with wind velocity U as the bifurcation parameter.

complex nonlinearities which could be responsible for the response surface behaviour. The deterministic bifurcation diagram in the absence of any random external excitations is given in Fig. 6.5(b).

The wind, when modelled as a Gaussian random process with three random variables ($d = 3$), the corresponding response surface shows sharp changes. This is shown in Fig. 6.6 with respect to one random variable ξ_1 for the ease of visualization, for a fixed value of mean wind. Expectedly, PCE cannot capture these multiple sharp changes. In fact it is seen to be smoothened out. The inadequacy of the PCE technique even with an increasing order, is demonstrated in the same figure. This is reproduced from the recent works of Devathi and Sarkar.[76] It is important to note here that the nonlinear nature of the stall behaviour is responsible for such response surface manifestation. Depending on the stall model, the qualitative nature of the irregularities could vary.

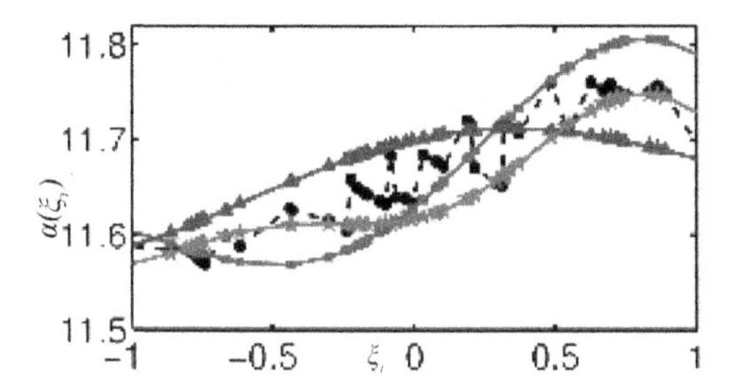

Fig. 6.6. Inadequacy of PCE (NISP) for the nonlinear aeroelastic system with gust: response α vs. ξ_1 (1-D response surfaces).[76] The broken dotted line is the actual response surface, firm lines with circle, triangle and star are for PCE expansion order $n = 15, 20, 30$ respectively.

The above two aeroelastic examples demonstrate that for sharply varying or irregular response surfaces, spectral tools like PCE cannot perform to their expected level of accuracy. In the rest of the chapter, we look for a viable alternative in another newly developed technique that is based on the principle of conservation of probability. It is called the Probability Density Evolution Method (PDEM).

4. Probability Density Evolution Method

A new class of uncertainty quantification tool proposed recently in the literature in nonlinear structural systems[78–85] is investigated here for aeroelastic applications. Despite several advances in the PCE techniques for various types of nonlinear and bifurcation behaviour, the issue of uneven response surface is still a challenge. As we demonstrated in the previous section, spectral tools have been found to be inadequate to capture such irregular response surfaces and corrective algorithms have been suggested only for a few limited types of discontinuous behaviour. In this regard, the merits of a relatively new method called Probability Density Evolution Method[81,86–88] is studied in the following.

4.1. *Formulation*

One of the most fundamental principles governing the evolution of a stochastic dynamical system[89] is the principle of preservation of probability, which, in a general sense can be stated as: if the random factors involved in a stochastic system are retained, the probability will be preserved in the evolution process of the system. Although this principle was known for long,[90] its physical interpretation and implementation has been done only recently.[80,91] The probability density evolution equations, like the Liouville equation, Dostupov-Pugachev equation and the Fokker-Planck equation have been rewritten from the principle of conservation which gave way to a new family of generalized density equations.

We consider an n-dimensional stochastic dynamical system in its state space form,

$$\dot{\vec{Y}} = A(\vec{Y}, t), \tag{6.14}$$

where, $\vec{Y}$ is the state vector of size $(2n \times 1)$ and $(\,\dot{}\,)$ denotes the derivative with respect to the time parameter t; let $\vec{Y_t}$ designate the state variables at time t and $\vec{Y_0}$ be the initial conditions. The state space governing Eq. (6.14) is a mapping from $\vec{Y_0}$ to $\vec{Y_t}$ and can be written as the following,

$$\vec{Y_t} = g(\vec{Y_0}, t), \quad \text{with } g \text{ as the mapping operator.} \tag{6.15}$$

Now, $\{\vec{Y_0} \in \Omega_0\}$ is random event in an arbitrary domain Ω_0 which changes to Ω_t at time t. The probability is preserved in the mapping of any arbitrary element event,

$$\int_{\Omega_0} p_{\vec{Y_0}}(\vec{y}, t_0)\, d\vec{y} = \int_{\Omega_t} p_{\vec{Y_t}}(\vec{y}, t)\, d\vec{y},$$

where, $p_{\vec{Y_t}}(\vec{y}, t)$ is the joint probability density function of $\vec{Y_t}$. As the above equation should hold true for any $(t + \Delta t)$ as well, one can generalize as the following,

$$\frac{D}{Dt}\int_{\Omega_t} p_{\vec{Y}_t}(\vec{y},t)\,d\vec{y} = 0, \quad \forall t. \tag{6.16}$$

Here, $\frac{D}{Dt}$ is the material derivative.

Now let us write the state space variables in terms of the generalized displacements and velocities so that one can follow the individual responses,

$$\vec{Y} = (\dot{\vec{X}}^T, \vec{X}^T)^T, \tag{6.17}$$

where, $\vec{X}$ = displacement vector of size $(n \times 1)$, $\dot{\vec{X}}$ = velocity vector of size $(n \times 1)$.

For a better understanding, a generic nonlinear dynamical system with the following governing equation is considered.

$$M\ddot{\vec{X}} + C\dot{\vec{X}} + K(\vec{X}) = F(t),$$

where, $F(t)$ is some forcing and $K(\vec{X})$ is a generic nonlinear function. This can be expressed in the following form.

$$\dot{\vec{Y}} = A(\boldsymbol{\xi}, \vec{Y}) + B(\boldsymbol{\xi}, t),$$

with,

$$A(\boldsymbol{\xi}, \vec{Y}) = \begin{Bmatrix} -M^{-1}C\dot{\vec{X}} - M^{-1}K \\ \dot{\vec{X}} \end{Bmatrix}, \quad B = \begin{Bmatrix} M^{-1}F \\ 0 \end{Bmatrix},$$

and, $\boldsymbol{\xi}_{d\times 1}$ is the total input random variables vector from the parameters (d_p) as well as force (d_f). Introducing an augmented state vector $\vec{\beta} = (\vec{Y}^T, \boldsymbol{\xi})^T$,

$$\dot{\vec{\beta}} = \begin{Bmatrix} \dot{\vec{Y}} \\ \dot{\boldsymbol{\xi}} \end{Bmatrix} = \begin{Bmatrix} A(\boldsymbol{\xi}, \vec{Y}) + B(\boldsymbol{\xi}, t) \\ 0 \end{Bmatrix} = G(\vec{Y}, \boldsymbol{\xi}, t). \tag{6.18}$$

Here, $\vec{\beta}$ is an augmented phase space. Specifying some random initial condition as, $(\vec{Y}^T(t_0), \boldsymbol{\xi}^T(t_0))^T = (\dot{\vec{X}}_0^T, \vec{X}_0^T, \boldsymbol{\xi}_0^T)$ and applying the principle of probability preservation (Eq. (6.16)) to the above system, one gets,

$$\frac{\partial}{\partial t}p_{\vec{Y}\boldsymbol{\xi}} + \sum_{j=1}^{2n+d} \dot{\beta}\frac{\partial}{\partial \beta_j}p_{\vec{Y}\boldsymbol{\xi}} = 0.$$

The material derivative has been elaborated in terms of temporal and spatial derivatives. This is in the Liouville's equation form.[89] The right-hand side is assumed to be '0' from the principle of preservation of probability. Li and Chen[92] argues that if one is interested in a single quality of interest Y_m of $\vec{Y}$ $(1 \leq m \leq 2n)$, then from of the definition of joint PDF,

$$p_{Y_m\boldsymbol{\xi}}(y_m, \boldsymbol{\xi}, t) = \int p_{\vec{Y}\boldsymbol{\xi}}(\vec{y}, \boldsymbol{\xi}, t)\, dy_1\, dy_2 \cdots dy_{m-1}\, dy_{m+1} \cdots dy_{2n},$$

one gets the following final expression after integrating the above form,

$$\frac{\partial p_{Y_m\boldsymbol{\xi}}}{\partial t} + \dot{Y}_m \frac{\partial}{\partial Y_m} p_{Y_m\boldsymbol{\xi}} = 0. \tag{6.19}$$

This equation is the final working form one needs, in terms of a single response dimension of interest.

4.2. *Solving the probability density evolution equation*

The numerical solution procedure of the PDEM technique requires a brief outline. The differential form of the conservation of PDF (Eq. 6.19) is the equation that needs to be solved for a particular stochastic dynamical system. This is called the probability density evolution equation (PDEE). In a simpler form, this can be written as,[93]

$$\frac{\partial P}{\partial t} + \dot{Y}_m \frac{\partial P}{\partial Y_m} = 0. \tag{6.20}$$

In the following, we keep up with the standard notations used in this chapter for PCE and Karhunen-Loève expansion and rewrite the PDEE that is required to be solved. It has been assumed that the time-variant joint probability density function of some response quantity Y_m and $\boldsymbol{\xi}$ (set of random variables from KL expansion) be denoted as $p(Y_m, \boldsymbol{\xi})$, referred to as P. Eq. (6.20) is the most generalized form of the PDEE. The initial and boundary conditions for Eq. (6.20) are given as the following,[81,86,87]

$$\begin{aligned} BC:\ & P \to 0 \quad \text{as} \quad \alpha, \boldsymbol{\xi} \to \pm\infty, \\ IC:\ & P|_{\tau=0} = \delta(\alpha - \alpha_0) p_{\boldsymbol{\xi}}, \end{aligned} \tag{6.21}$$

where, $p_{\boldsymbol{\xi}}$ is the joint probability density function of the elements of the vector $\boldsymbol{\xi}$, δ indicates the Dirac-delta function, α_0 is the initial conditions for the response and its derivative (aeroelastic response). Let $\wp$ denote the time-variant marginal density function of Y_m which can be obtained from P using the following d-dimensional integral,

$$\wp = \int_{-\infty}^{\infty} \cdots \int_{-\infty}^{\infty} P d\xi_1 d\xi_2 \cdots d\xi_d. \tag{6.22}$$

Here, d is the total number of terms employed in the Karhunen-Loève expansion. The above integral can be approximated numerically using a quadrature rule. To ensure optimal convergence with the minimal number of quadrature points, a Gauss-Hermite quadrature has to be employed for input Gaussian distribution and Eq. (6.22) can be rewritten in the following form,

$$\wp = \sum_{i=1}^{n^d} \mathfrak{f}_i w_i, \tag{6.23}$$

where, n is the discretization order of the quadrature and same as that used in the PCE, $\mathfrak{f}_i$ is the value of the function $\mathfrak{f}$ at the ith quadrature point and w_i is the corresponding quadrature weight. The quadrature points are nothing but the roots of the multi-dimensional Hermite polynomials, $\Phi(\boldsymbol{\xi}) = \phi(\xi_1) \otimes \phi(\xi_2) \otimes \cdots \otimes \phi(\xi_d)$ of the chosen order.[45] Equation (6.20) is a scalar, hyperbolic partial differential equation whose solution strategies are available in the literature.[94–96]

The main challenges of the numerical scheme are to choose the collocation/representative points and to choose the scheme for solving the partial differential equation. However, this is done for probability and not for the response itself. The solution procedure in PDEM would involve a number of deterministic runs (point evolution at collocation points) which requires partitioning of the random domain in a multi-variate case and also numerically solving the partial differential equations.[83,92] The Lax-Wendroff scheme[97,98] is a commonly used technique which is a second order accurate finite difference scheme to solve for partial differential equations. However, it is a dispersion type of scheme and cannot preserve the non-negative nature of the PDFs. In order to prevent spurious values of the PDFs, imposition of flux limiters[99] has been used in the recent work of one of the authors.[76] Using that, a modified Lax-Wendroff algorithm which is a total variation diminishing (TVD) scheme[81,86] was suggested. This ensures non-negativity of the PDFs and also the spurious oscillations in the PDF behaviour. The solution of a hyperbolic partial differential equation is generally involved with steep gradients in variable values, such as encountered in shocks. A non-monotone scheme can engender in spurious oscillations in the shock vicinity. A TVD scheme, which is a monotone scheme, can keep such oscillations under control. Further, the TVD scheme, by means of employing flux limiters, also ensures that the PDFs are strictly positive.[94–96,100]

In order to complete the ongoing discussion on the numerical procedure, we present the spatial and temporal discretization steps with the help of the of PDEE form used in this section.[45] Let $P_{k,l}$ indicate the value of P, at the lattice point $\{y_k, t_l\}$. The finite difference scheme for solving Eq. (6.20) can then be written as,[96]

$$\begin{aligned} P_{k,l+1} = P_{k,l} - r_L\Big(&\tfrac{1}{2}(\dot{Y}m_{l+1} + |\dot{Y}m_{l+1}|)(P_{k,l} - P_{k-1,l}) \\ &+ \tfrac{1}{2}(\dot{Y}m_{l+1} - |\dot{Y}m_{l+1}|)(P_{k+1,l} - P_{k,l})\Big) \\ &- \tfrac{1}{2}(1 - |r_L\dot{Y}m_{l+1}|)|r_L\dot{Y}m_{l+1}|(\mathfrak{F}(r^+_{k+\frac{1}{2}}, r^-_{k+\frac{1}{2}})(P_{k+1,l} \\ &- P_{k,l}) - \mathfrak{F}(r^+_{k-\frac{1}{2}}, r^-_{k-\frac{1}{2}})(P_{k,l} - P_{k-1,l})), \end{aligned} \tag{6.24}$$

where r_L is the ratio of the temporal and the spatial discretization steps used in the finite difference scheme, $r^{\pm}_{k+\frac{1}{2}}$ and $r^{\pm}_{k-\frac{1}{2}}$ are solution gradients defined as,

$$\begin{aligned} r^+_{k+\frac{1}{2}} &= (P_{k+2,l} - P_{k+1,l})/(P_{k+1,l} - P_{k,l}), \\ r^-_{k+\frac{1}{2}} &= (P_{k,l} - P_{k-1,l})/(P_{k+1,l} - P_{k,l}), \\ r^+_{k-\frac{1}{2}} &= (P_{k+1,l} - P_{k,l})/(P_{k,l} - P_{k-1,l}), \\ r^-_{k-\frac{1}{2}} &= (P_{k-1,l} - P_{k-2,l})/(P_{k,l} - P_{k-1,l}). \end{aligned} \tag{6.25}$$

$\mathfrak{F}$ in Eq. (6.24) is the flux limiter term. The value of r_L should be taken such that the Courant-Friedrichs-Lewy (CFL) number ≤ 1 at every time step.

5. Application of PDEM to Aeroelastic Problems

The PDEM tool is implemented to the two aeroelastic systems we have mentioned earlier in this chapter, which show uneven response surfaces. As we have seen, the PCE approaches cannot be quite suitable in such situations and they would smoothen out the response around the sharp changes. Qualitatively the response surface behaviour of the subcritical nonlinear system is similar to the stall system under random gust. The inadequacy of the PCE tool shown in the response surface behaviour would eventually translate to incorrect response statistics. Hence, this is a situation in which it is more efficient to follow the time evolution of the response PDF itself using tools like PDEM. For the subcritical system, the PDEM results are shown in Fig. 6.7, in which PDEM matches with MCS with good accuracy. As PDEM computes the PDF of the response directly it cannot be compared on the response surface plot. It is worth noting here that, there have been some response surface interpolation based techniques developed specifically for subcritical systems. However, the application of such tools are confined to specific nonlinearities alone[60,66,67,101] and would lose their efficiency when the response surface undergoes many sharp irregularities or multiple changes in rapid succession. On the other hand, PDEM can be developed as an universal approach irrespective of the type and complexity of the unevenness. We test this further in the next problem of stall aeroelastic system. The nonlinearity in the stall model has been responsible for a severely uneven response surface, which was difficult to model using PCE.

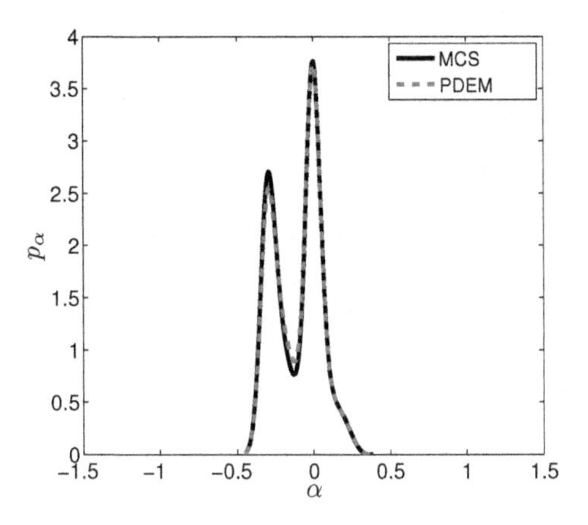

Fig. 6.7. Probability density function by PDEM and MCS for the aeroelastic system with subcritical behaviour; $U = 6.2$, $\bar{\beta} = -3$ is a uniformly distributed random variable.

In this case, the PDF evolution surfaces with time have been used to highlight the differences between PCE and PDEM. As before, response surface plots cannot be presented for PDEM. This comparison is presented in Figs. 6.8(a)–6.8(c) for MCS, PCE and PDEM respectively. It can be seen that the mountain-valley structure, clearly seen in Figs. 6.8(a) and 6.8(c) is not captured by PCE. The PDF pattern by PCE in fact predict the qualitative trend to be single modal which is not correct for the present situation. This inadequacy of PCE which was visible in the response surface plots (Fig. 6.6) could be attributed to its inability to adjust to the response surface behaviour.

Finally, we conclude that PDEM may be a better choice in situations involving sharp response gradients in the random domain, in which PCE based approaches which rely on polynomial expansions fail. In that sense,

PDEM could emerge as a promising tool for UQ in aeroelastic systems involving complex stochastic bifurcation behaviour. It should be noted that the failure of PCE shown in the present work have been based on the use of global polynomials. On the other hand, multi-element based PCE approaches[36,38] with local support that are proposed in the literature (discussed in Sec. 2.1) can work in a piece-wise manner and can be implemented to tackle sharp changes. However, for a large number of irregularities as encountered presently (and in a multi-dimensional space), they are also likely to become ineffective.

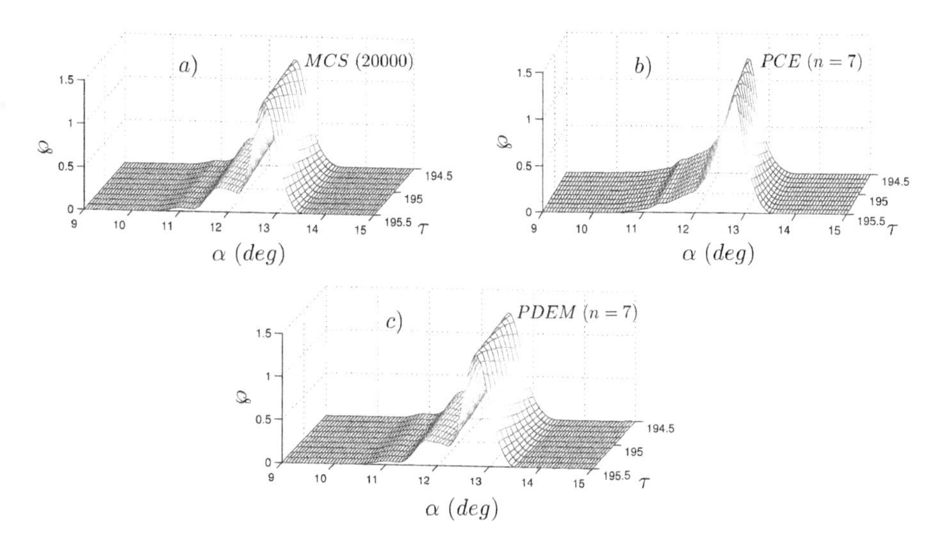

Fig. 6.8. Validation study. (a) MCS method; (b) PCE method; (c) PDEM.[45]

6. Conclusions

The main focus of the present article has been to highlight the scope of a recently proposed tool called the Probability Density Evolution method in nonlinear aeroelastic problems. It has been shown that PDEM has certain distinct advantages over the traditional PCE method. However, it could also suffer from certain disadvantages, especially in terms of computational efforts. In addition to solving the stochastic differential equation governing the evolution of the random response at the various collocation points, it is required to solve the hyperbolic probability conservation law at each of these collocation points. This implies that the runtime would be, at least, twice that required for the PCE method. Also, the evolution equation has to be derived separately for each response degree of freedom and the finite difference grid also has to be constructed independently. This could limit its application to very large order problems. However, for very large random dimensional problems (in industrial problems or problems with real-life applications), use of PCE and its variants becomes a tedious task as well. For such cases, it would probably still be computationally more viable to use sampling based tools like MCS or intelligent MCS. Nevertheless, more efficient adaptive variant of PDEM could be considered as an interesting future work direction.

Further, PDEM has a lot of similarity with the well-established Fokker Planck equation approach. In the latter technique, one deals with both the convective and diffusive fluxes from the corresponding Ito equation form, unlike PDEM which deals with convective fluxes alone. The Fokker Planck equation can be solved provided the coefficients of the flux terms are available in analytical forms which may be difficult for complex nonlinear systems. In that sense, PDEM can offer a distinct advantage as in this case, the coefficients of the flux term can be calculated numerically as well. Hence, PDEM looks more promising for applications in complex nonlinear problems.

References

1. N. Wiener, The homogeneous chaos, *Amer. J. Math.* **60**, 897–936 (1938).
2. R. Ghanem and P. D. Spanos, *Stochastic Finite Elements: A Spectral Approach.* Springer-Verlag, New York (1991).
3. G. E. Karniadakis and D. Lucor, Predictability and uncertainty in flow-structures interactions, *Eur. J. Mech. B Fluids* **23**, 41–49 (2004).
4. G. Lin, C. H. Su and G. Karniadakis, Predicting shock dynamics in the presence of uncertainties, *J. Comput. Phys.* **217**(1), 260–276 (2006).
5. L. Bruno, C. Canuto and D. Fransos, Stochastic aerodynamics and aeroelasticity of a flat plate via generalized polynomial chaos, *J. Fluids and Struct.* **25**, 1158–1176 (2009).
6. A. Manan and J. Cooper, Design of composite wings including uncertainties: a probabilistic approach, *J. Aircraft* **46**, 601–607 (2009).
7. C. Scarth, J. E. Cooper, P. M. Weaver and G. Silva, Uncertainty quantification of aeroelastic stability of composite plate wings using lamination parameters, *Composite Struct.* **116**, 84–93 (2014).
8. S. Sarkar, S. Gupta and I. Rychlik, Wiener chaos expansions for estimating rain-flow fatigue damage in randomly vibrating structures with uncertain parameters, *Probab. Eng. Mech.* **26**, 387–398 (2010).

9. D. Xiu and G. E. Karniadakis, The Wiener-Askey polynomial chaos for stochastic differential equations, *SIAM J. Sci. Comput.* **24**, 619–644 (200).
10. R. Askey and J. Wilson, Some basic hypergeometric polynomials that generalize jacobi polynomials, *Mem. Amer. Math. Soc.* **54**, 319 (1964).
11. R. Ghanem, Stochastic finite elements with multiple random non-Gaussian properties, *J. Eng. Mech.* **125**(1), 26–40 (1999).
12. D. Xiu and G. Karniadakis, Modeling uncertainty in flow simulations via generalized polynomial chaos, *J. Comput. Phys.* **187**(1), 137–167 (2003).
13. D. Xiu and G. E. Karniadakis, A new stochastic approach to transient heat conduction modeling with uncertainty, *Int. J. Heat Mass Transf.* **46**(24), 4681–4693 (2003).
14. G. Lin, X. Wan, C. Su and G. E. Karniadakis, Stochastic computational fluid mechanics, *Comput. Sci. Eng.* **9**, 21–29 (2007).
15. J. A. S. Witteveen, S. Loeven, A. Sarkar and H. Bijl, Probabilistic collocation for period-1 limit cycle oscillations, *J. Sound Vibrations* **311**, 421–439 (2008).
16. G. J. A. Loeven, J. A. S. Witteveen and H. Bijl, Probabilistic collocation: an efficient non-intrusive approach for arbitrarily distributed parametric uncertainties, *AIAA 2007-317, 45th AIAA Aerospace Sciences Meeting and Exhibit*, Reno, NV (2007).
17. A. Loeven, Efficient uncertainty quantification in computational fluid dynamics, Ph.D. thesis, Technical University of Delft, Delft, The Netherlands (2010).
18. R. W. Walters, Towards stochastic fluid mechanics via polynomial chaos, *AIAA*-2003-0413, *41st AIAA Aerospace Sciences Meeting and Exhibit*, Reno, NV (2003).
19. S. Hosder, R. W. Walters and R. Perez, A non-intrusive polynomial chaos method for uncertainty propagation in CFD simulations, *AIAA*-2006-891, *44th AIAA Aerospace Sciences Meeting and Exhibit*, Reno, NV (2006).
20. S. Oladyshkin, H. Class, R. Helmig and W. Nowak, An integrative approach to robust design and probabilistic risk assessment for co2 storage in geological formations, *Comput. Geosci.* **15**, 565–577 (2011).
21. C. L. Pettit and P. S. Beran, Spectral and multiresolution Wiener expansions of oscillatory stochastic process, *J. Sound Vibration* **294**, 752–779 (2006).
22. A. Desai and S. Sarkar, Analysis of a nonlinear aeroelastic system with parametric uncertainties using polynomial chaos expansion, *Math. Probl. Eng.* **2010**(379472), 1–21 (2010).
23. A. Desai, J. A. S. Witteveen and S. Sarkar, Uncertainty quantification of a nonlinear aeroelastic system using polynomial chaos expansion with constant phase interpolation, *J. Vibration Acoust.* **135**(5), 051034 (2013).
24. A. H. Stroud and D. Secrest, *Gaussian Quadrature Formulas.* Prentice-Hall, Inc., Englewood Cliffs, NJ, USA (1966).
25. A. Alexanderian, A brief note on the Karhunen-Loève expansion, On line article: http://users.ices.utexas.edu/ alen/articles/KL.pdf (2013).

26. R. H. Cameron and W. T. Martin, The orthogonal development of non-linear functionals in series of Fourier–Hermite functionals, *Ann. of Math.* 385–392 (1947).
27. D. Xiu and G. E. Karniadakis, Modeling uncertainty in steady state diffusion problems via generalized polynomial chaos, *Comput. Methods Appl. Mech. Engrg.* **191**(43), 4927–4948 (2002).
28. X. Wan and G. Karniadakis, Beyond Wiener–Askey expansions: handling arbitrary PDFs, *J. Sci. Comput.* **27**(1–3), 455–464 (2006).
29. J. A. S. Witteveen, S. Sarkar and H. Bijl, Modeling physical uncertainties in dynamic stall induced fluid–structure interaction of turbine blades using arbitrary polynomial chaos, *Comput. Struct.* **85**(11), 866–878 (2007).
30. M. T. Reagan, H. N. Najm, R. Ghanem and O. M. Knio, Uncertainty quantification in reacting-flow simulations through non-intrusive spectral projection, *Combust. Flame* **132**(3), 545–555 (2003).
31. O. M. Le Matre, O. P. Knio, H. N. Najm and R. G. Ghanem, A stochastic projection method for fluid flow: I. Basic formulation, *J. Comput. Phys.* **173**(2), 481–511 (2001).
32. G. H. Golub and J. H. Welsch, Calculation of Gauss quadrature rules, *Math. Comput.* **23**(106), 221–230 (1969).
33. J. Ma, V. Rokhlin and S. Wandzura, Generalized Gaussian quadrature rules for systems of arbitrary functions, *SIAM J. Numer. Anal.* **33**(3), 971–996 (1996).
34. C. Soize and R. Ghanem, Physical systems with random uncertainties: chaos representations with arbitrary probability measure, *SIAM J. Sci. Comput.* **26**(2), 395–410 (2004).
35. T. Gerstner and M. Griebel, Numerical integration using sparse grids, *Numer. Algorithms* **18**, 209–232 (1998).
36. J. A. S. Witteveen, A. Loeven and H. Bijl, An adaptive stochastic finite elements approach based on Newton–Cotes quadrature in simplex elements, *Comput. Fluids* **38**(6), 1270–1288 (2009).
37. J. A. S. Witteveen and H. Bijl, An alternative unsteady adaptive stochastic finite elements formulation based on interpolation at constant phase, *Comput. Methods Appl. Mech. Engrg.* **198**(3), 578–591 (2008).
38. J. C. Chassaing, D. Lucor and J. Trégon, Stochastic nonlinear aeroelastic analysis of a supersonic lifting surface using an adaptive spectral method, *J. Sound Vibration* **331**(2), 394–411 (2012).
39. H. J. Bungartz and M. Griebel, Sparse grids, *Acta Numerica* **13**, 147–269 (2004).
40. J. Garcke, M. Griebel and M. Thess, Data mining with sparse grids, *Computing* **67**(3), 225–253 (2001).
41. B. Ganapathysubramanian and N. Zabaras, Sparse grid collocation schemes for stochastic natural convection problems, *J. Comput. Phys.* **225**(1), 652–685 (2007).
42. X. Ma and N. Zabaras, An adaptive hierarchical sparse grid collocation algorithm for the solution of stochastic differential equations, *J. Comput. Phys.* **228**(8), 3084–3113 (2009).

43. H. J. Bungartz and S. Dirnstorfer, Multivariate quadrature on adaptive sparse grids, *Computing* **71**(1), 89–114 (2003).
44. H. Devathi and S. Sarkar, Quantification of parametric uncertainties in a dynamic stall induced aeroelastic system using an adaptive sparse grid quadrature method, submitted, under review (2016).
45. H. Devathi, A stall induced fluid-structure interaction system under the influence of parametric uncertainties and random gust, Master's thesis, Indian Institute of Technology Madras, Chennai, India (2014).
46. R. A. Ibrahim, D. M. Beloiu and C. L. Pettit, Influence of joint relaxation on deterministic and stochastic panel flutter, *AIAA J.* **43**(7), 1444–1454 (2005).
47. P. J. Attar, E. H. Dowell and J. R. White, Modeling delta wing limit-cycle oscillations using a high-fidelity structural model, *J. Aircraft* **42**(5), 1209–1217 (2005).
48. S. Choi and N. Namachchivaya, Stochastic dynamics of a nonlinear aeroelastic system, *AIAA J.* **44**, 1921–1931 (2006).
49. N. S. Namachchivaya and J. Prussing, Almost-sure asymptotic stability of rotor blades flapping motion in forward flight in turbulent flow, *Probab. Eng. Mech.* **6**(1), 2–9 (1991).
50. N. S. Namachchivaya, Stochastic bifurcation, *Appl. Math. Comput.* **38**, 101–159 (1990).
51. D. Poirel and S. J. Price, Response probability structure of a structurally nonlinear fluttering airfoil in turbulent flow, *Probab. Eng. Mech.* **18**(2), 185–202 (2003).
52. D. Poirel and S. J. Price, Bifurcation characteristics of a two-dimensional structurally non-linear airfoil in turbulent flow, *Nonlinear Dynam.* **48**(4), 423–435 (2007).
53. D. Poirel and S. J. Price, Random binary (coalescence) flutter of a two-dimensional linear airfoil, *J. Fluids Struct.* **18**, 23–42 (2003).
54. D. C. Poirel and S. J. Price, Post-instability behavior of a structurally nonlinear airfoil in longitudinal turbulence, *J. Aircraft* **34**(5), 619–626 (1997).
55. A. N. Desai, Analysis of a nonlinear aeroelastic system with parametric uncertainties using polynomial chaos expansion, Master's thesis, Indian Institute of Technology Madras, Chennai, India (2011).
56. V. Nair, G. Thampi and R. Sujith, Intermittency route to thermoacoustic instability in turbulent combustors, *J. Fluid Mech.* **756**, 470–487 (2014).
57. V. Nair and R. I. Sujith, Multifractality in combustion noise: predicting an impending combustion instability, *J. Fluid Mech.* **747**, 635–655 (2014).
58. J. Venkatramani, V. Nair, R. I. Sujith, S. Gupta and S. Sarkar, Precursors to flutter instability by an intermittency route, *J. Fluids Struct.* **61**, 376–391 (2016).
59. W. Horsthemke, *Noise Induced Transitions*. Springer (1984).
60. M. Patel, V. Nair and S. Sarkar, Uncertainty quantification of an airfoil control surface flutter system, In: eds. G. Deodatis, B. R. Ellingwood and D. M. Frangopol, *Safety, Reliability, Risk and Life-Cycle Performance of Structures and Infrastructures*, Chapter 750 (2013).

61. B. H. K. Lee, L. Y. Jiang and Y. S. Wong, Flutter of an airfoil with a cubic restoring force, *J. Fluids Struct.* **13**(1), 75–101 (1999).
62. S. Venkatesh, S. Sarkar and I. Rychlik, Uncertainties in blade flutter damage prediction under random gust, *Probab. Eng. Mech.* **36**, 45–55 (2014).
63. Y. Fung, *An Introduction to the Theory of Aeroelasticity*. Courier Dover Publications (2002).
64. H. Alighanbari, Flutter analysis and chaotic response of an airfoil accounting for structural nonlinearities, Ph.D. thesis, McGill University (1995).
65. S. H. Strogatz, *Nonlinear Dynamics and Chaos: With Applications to Physics, Biology and Chemistry*. Perseus Publishing (2001).
66. V. Nair, S. Sarkar and R. I. Sujith, Uncertainty quantification of subcritical bifurcations, *Probab. Eng. Mech.* **34**, 177–188 (2013).
67. P. S. Beran, C. L. Pettit and D. R. Millman, Uncertainty quantification of limit-cycle oscillations, *J. Comput. Phys.* **217**(1), 217–247 (2006).
68. V. Nair, S. Sarkar and R. I. Sujith, Uncertainty quantification of subcritical bifurcation in a rijke tube, *AIAA*-2010-3858. *16th AIAA/CEAS Aeroacoustic Conference* (2010).
69. R. Lefever and J. W. Turner, Sensitivity of a Hopf bifurcation to multiplicative colored noise, *Phys. Rev. Lett.* **56**(16), 1631 (1986).
70. D. R. Millman, Quantifying initial conditions and parametric uncertainties in a nonlinear aeroelastic system with an efficient stochastic algorithm, Ph.D. thesis, Air force Institute of Technology, Wright-Patterson Airforce Base, Ohio, USA (September, 2004).
71. D. Peters, Towards a unified lift model for use in rotor blade stability analyses, *J. Amer. Helicopter Soc.* **30**, 32–42 (1985).
72. C. Tran and T. Petot, Semi-empirical model for the dynamic stall of airfoils in view of the application to the calculation of responses of a helicopter blade in forward flight, *Vertica* **5**, 35–53 (1981).
73. J. G. Leishman and T. S. Beddoes, A semi-empirical model for dynamic stall, *J. Amer. Helicopter Soc.* **34**(3), 3–17 (1989).
74. C. E. Rasmussen and C. K. I. Williams, Gaussian processes for machine learning (2006).
75. K. E. Kaasen, Time domain model representations of standard wind gust spectra, In *Proceedings of the International Offshore and Polar Engineering Conference* (1999).
76. H. Devathi and S. Sarkar, Study of a stall induced dynamical system under gust using the probability density evolution technique, *Comput. Struct.* **162**, 38–47 (2016).
77. S. Price and G. Fragiskatos, Nonlinear aeroelastic response of a two-degree-of-freedom airfoil oscillating in dynamic stall, In *Proceedings of the 7th International Conference on Flow Induced Vibration*, Vol. 2000, pp. 437–444 (2000).
78. J. B. Chen and J. Li, Dynamic response and reliability analysis of non-linear stochastic structures, *Probab. Eng. Mech.* **20**, 33–44 (2005).
79. J. Li and J. Chen, The number theoretical method in response analysis of nonlinear stochastic structures, *Comput. Mech.* **39**(6), 693–708 (2007).

80. J. Li, J. B. Chen and W. Fan, The equivalent extreme-value event and evaluation of the structural system reliability, *Struct. Saf.* **29**, 112–131 (2007).
81. J. Li and J. B. Chen, Probability density evolution method for dynamic response analysis of structures with uncertain parameters, *Comput. Mech.* **34**(5), 400–409 (2004).
82. J. B. Chen and R. Ghanem, Partition of the probability-assigned space in probability density evolution analysis of nonlinear stochastic structures, *Probab. Eng. Mech.* **24**, 27–42 (2009).
83. J. B. Chen and J. Li, Strategy for selecting representative points via tangent spheres in the probability density evolution method, *Int. J. Numer. Methods Eng.* **74**, 1988–2014 (2008).
84. J. Li, Q. Yan and J. B. Chen, Stochastic modeling of engineering dynamic excitations for stochastic dynamics of structures, *Probab. Eng. Mech.* **27**, 19–28 (2012).
85. J. Li, J. Chen, W. Suna and Y. Peng, Advances of the probability density evolution method for nonlinear stochastic systems, *Probab. Eng. Mech.* **28**, 132–142 (2012).
86. J. Li and J. B. Chen, The probability density evolution method for analysis of dynamic nonlinear response of stochastic structures, *Acta Mech. Sinica* **6**, 008 (2003).
87. J. Li and J. B. Chen, The probability density evolution method for dynamic response analysis of non-linear stochastic structures, *Int. J. Numer. Methods Eng.* **65**(6), 882–903 (2006).
88. J. Li and J. Chen, The principle of preservation of probability and the generalized density evolution equation, *Struct. Saf.* **30**(1), 65–77 (2008).
89. R. Syski, Stochastic differential equations. In: ed. T. L. Saaty, *Modern Nonlinear Equations*, Chapter 8. McGraw-Hill, New York (1967).
90. B. G. Dostupov and V. S. Pugachev, The equation for the integral of a system of ordinary differential equations containing random parameters, *Automati. Telemekh.* **18**, 620–630 (1957).
91. J. Li and J. B. Chen, *Stochastic Dynamics of Structures.* Wiley (2009).
92. J. B. Chen and J. Li, A note on the principle of preservation of probability and probability density evolution equation, *Probab. Eng. Mech.* **24** (2009).
93. R. Zwanzig, *Nonequilibrium Statistical Mechanics.* Oxford University Press (2001).
94. H. C. Yee, Construction of explicit and implicit symmetric TVD schemes and their applications, *J. Comput. Phys.* **68**(1), 151–179 (1987).
95. P. K. Sweby, High resolution schemes using flux limiters for hyperbolic conservation laws, *SIAM J. Numer. Anal.* **21**(5), 995–1011 (1984).
96. M. Y. Shen, Z. B. Zhang and X. L. Niu, Some advances in study of high order accuracy and high resolution finite difference schemes, In *New Advances in Computational Fluid Dynamics.* pp. 111–145 (2001). Higher Education Press, Beijing.
97. J. Gerrity and P. Joseph, A note on the computational stability of the two-step Lax-Wendroff form of the advection equation, *Monthly Weather Rev.* **100**, 72–73 (1972).

98. J. Qiu, A numerical comparison of the Lax–Wendroff discontinuous Galerkin method based on different numerical fluxes, *J. Sci. Comput.* **30**, 345–367 (2007).
99. Z. Xu, Parameterized maximum principle preserving flux limiters for high order schemes solving hyperbolic conservation laws: one dimensional scalar problem, *Math. Comput.* **83**, 2213–2238 (2013).
100. A. Harten, High resolution schemes for hyperbolic conservation laws, *J. Comput. Phys.* **49**(3), 357–393 (1983).
101. D. R. Millman, P. I. King, R. C. Maple and P. Beran, Predicting uncertainty propagation in a highly nonlinear system with a stochastic projection method, *AIAA* 2004-1613, *45th AIAA/ASCE/AHS/ASC Structures, Structural Dynamics, and Materials Conference* (April, 2004).

Index

A

acoustic noise, 1
aeroacoustics, vii, 1

B

Bayesian, 51, 134–136
bifurcations, 27, 151–153, 156–162, 166–167, 173
bifurcation (D-bifurcation), 158
bifurcation (P-bifurcation), 158

C

cavitation, 91–95, 107, 109, 113–115
cavitation phenomenon, vii
collocation, 152, 154–157, 165
conservation equations, 102

D

DEM, 95, 98
differential equations, 121, 131

E

equation of state (EOS), 40, 99–101, 104–108, 111–113
error bars, 23, 28

F

failure, 125–131, 160

G

Gaussian processes, 122, 131–132

I

inverse problems, vii, 124, 134, 136

K

Karhunen-Loève (KL), 122–123, 164–165
Karhunen-Loève expansion, 154, 164, 169–170
KL approximation, 123

L

Large-eddy simulation (LES), 2–3, 5–6, 8, 10, 12–14, 16, 18, 20–21, 27, 45
LES-SC, 27

M

Monte Carlo (MC) simulations, 1, 7, 28, 38, 58, 119, 126, 135, 138, 151–152, 154
multiplicative, 153, 162

N

Navier, 12
Navier-Stokes (RANS), 1, 7
noise, 1–3, 6, 9, 18, 24–26, 28, 45, 53–54, 56, 73, 91, 134
non-Gaussian processes, 123, 131–132
non-intrusive spectral projection (NISP), 154–155

nonlinear, 40, 54, 86, 124, 134, 138, 151–152, 156, 158, 160, 166–168, 171, 173–174
nonlinear effects, 39
nonlinearity, 135–136, 138, 158, 159, 163, 166, 172
nonlinearization, 131

P

P-bifurcation, 158
parametric excitations, 127, 153
parametric randomness, 156, 158
point-collocation, 63–66, 87
polynomial chaos, vii, 50, 63–67, 87, 94, 106, 122, 151–155, 161
Probability Density Evolution Method (PDEM), 153, 166, 169–174

R

RANS, 2–3, 5–7, 10–13, 16, 18, 20–22, 26–28
RANS-based, 20
RANS-SC, 27
RANS-UQ, 27
reliability, 119–120, 125–128, 130–133
response surface, 17, 66, 71–72, 82, 136, 151, 154, 157, 160–163, 166–167, 171–172

S

self-noise, 1–2, 26
sensitivity, 33
sensitivity analysis, 33–34, 38, 48, 51–52, 57, 69, 85–87, 94, 136–137
Sobol index, 69–71, 73, 82, 86–87, 137
sparse grid, 125, 152, 156–157
stochastic collocation, 1, 7, 13–14, 125
stochastic finite element, 119, 121
surrogate model, 38, 48, 52, 63–64, 66–67, 71–72, 74, 81–82, 85, 136
surrogate modelling, 64–65, 87

U

uncertainty quantification
  PC, 7
  sensitivity analysis, 7
URANS, 23

W

white noise, 131

Printed in the United States
By Bookmasters